49. 95

THE GIS GUIDE TO PUBLIC DOMAIN DATA

JOSEPH J. KERSKI
JILL CLARK

Esri Press
REDLANDS|CALIFORNIA

Esri Press, 380 New York Street, Redlands, California 92373-8100

16 15 14 13 12   1 2 3 4 5 6 7 8 9 10

Library of Congress Cataloging-in-Publication Data
Kerski, Joseph J.
    The GIS guide to public domain data / Joseph J. Kerski, Jill Clark.
        p. cm.
    Includes bibliographical references and index.
    ISBN 978-1-58948-244-9 (pbk. : alk. paper)
1. Geospatial data. 2. Public domain--Geographic information systems. I. Clark, Jill. 1962- II. Title.
    G70.217.G46K47 2012
    025.06'910285--dc23   2011051646

Ask for Esri Press titles at your local bookstore or order by calling 800-447-9778, or shop online at esri.com/esripress. Outside the United States, contact your local Esri distributor or shop online at eurospanbookstore.com/esri.

Esri Press titles are distributed to the trade by the following:

In North America:

Ingram Publisher Services
Toll-free telephone: 800-648-3104
Toll-free fax: 800-838-1149
E-mail: customerservice@ingrampublisherservices.com

In the United Kingdom, Europe, Middle East and Africa, Asia, and Australia:

Eurospan Group
3 Henrietta Street
London WC2E 8LU
United Kingdom
Telephone: 44(0) 1767 604972
Fax: 44(0) 1767 601640
E-mail: eurospan@turpin-distribution.com

Contents

Foreword

We are said to live in the age of the exaflood, when accessing the Internet is sometimes likened to drinking from a fire hose. It's not so long ago, well within my adult memory, that people were impressed by kilobytes. By the late 1980s there were floppy disks with megabytes. Then, in quick succession came the gigabytes of DVDs and thumb drives, the terabytes of external hard drives, and the petabytes that Google distributes every day from its servers. An exabyte is 10^{18} bytes, or a billion billion billion bytes, and it surely won't be long before we have devices capable of storing an exabyte that fit into the palm of a human hand.

In this wonderland of publicly available data, it's sometimes hard to believe that the right bytes can be hard to find, that looking for useful data can be like searching for a sometimes nonexistent needle in an enormous haystack. Yet, the vast majority of the data captured by the remote-sensing satellites that circle the earth are never actually examined in any detail. Finding the right data to solve a problem, and determining whether those data are fit for use, can be a challenging, drawn-out, and frustrating task. Data can be too coarse or noisy, too poorly documented, subject to restrictive terms of use, or simply not available for the right area at the right time.

Geospatial technologies have evolved over the past four decades to provide tremendous power in support of problem solving and are used to answer questions such as

Where is the best place to put my manufacturing plant?

How should I schedule my fleet of trucks to deliver packages at minimum cost?

What timber stands should I make available for cutting in the coming season?

What should I do to preserve adequate habitat for an endangered species?

Where should I locate the next retail store in my expanding chain?

Many, if not most, of the datasets needed to solve problems such as these exist in the public domain. But the steps of searching for them, assessing their appropriateness and fitness for use, dealing with the conditions imposed by their custodians, and accessing them for analysis can be cumbersome and complex. This book provides a systematic guide to the process, addressing issues that a user is likely to confront.

Conditions of access to geospatial data vary markedly across the world. In the United States, the Founding Fathers were adamant that the data acquired by the national government should be freely available to all—a principle that today prohibits federal agencies from charging more than the cost of dataset reproduction—not the personal property of the monarch, as in many European counties. In the United Kingdom, the doctrine of Crown Copyright is still largely intact. In other parts of the world, access to geospatial data is often restricted in the name of national security. Since the events of September 11, 2001, public access has been denied even in the United States to many types of government data that were previously freely available. Geospatial data have always had enormous strategic value, so it is scarcely surprising that national security is often cited to deny public access.

Against this, however, stands the argument that information should circulate freely in a modern society—that a democratic society is also an open society. This is the argument, often passionately made, that led to the establishment of WikiLeaks, the open-source software movement, and numerous data liberation initiatives around the world.

Over the past two decades, substantial investments in geospatial data infrastructures have facilitated the storage, discovery, access, and use of geospatial data. They include standards for data documentation or metadata, formats and specifications that facilitate interoperability across platforms, massive stores of online data with well-designed methods for search, and standards that allow users to access and process data remotely without the need for download. These arrangements are notably more advanced in the geospatial world than in many others and have been widely emulated in other domains. We can access and make use of geospatial data as never before. At the same time, the data available have been increasing exponentially, with no end in sight.

This book fills a very big gap in the literature of GIS and brings together for the first time discussions of issues that users of public domain data are likely to confront. It will prove useful to GIS practitioners in any area of GIS application, including students anxious to learn the skills needed to become GIS practitioners and data producers who want their data to be as useful as possible. The problems that GIS seeks to solve, whether in resource management, planning, utilities, logistics, or many other application domains, will all be a little closer to solution for readers who assimilate the contents of this book.

Michael F. Goodchild

Professor of geography at the University of California, Santa Barbara, and director of UCSB's Center for Spatial Studies.

Acknowledgments

We would like to thank all who have helped us compile the many references used in this book, including Sue Willetts, senior library assistant at the Institute of Classical Studies Library and the Joint Library of the Hellenic & Roman Societies in London, who suggested the two Roman examples. Thanks to the following for their assistance in compiling the case studies in chapter 7: Dominic Habron, senior GIS analyst at SEPA; Dr. Roselyne Lacaze at the Earth Observation Department, Center for the Study of the Biosphere from Space (CESBIO); and Matthew Steil, an associate with the People & Ecosytems program, World Resources Institute.

We also are grateful to the many people who either provided, or helped us to ascertain, permission to use the spatial data and numerous online resources cited in this book, including Minna Pyhälä; at The Helsinki Commission; Gary Meadows at Ordnance Survey; André Luiz at Imagem Geosistemas (Esri Distributor in Brazil); Stephen Dembner at the Food and Agriculture Organization of the United Nations; Elizabeth Colvard at the US Geological Survey; and Gerald Esch at the National Oceanic and Atmospheric Administration.

We wish to express our gratitude to the staff of Esri Press for support throughout this project.

Lastly, we want to acknowledge all the GIS practitioners who, every day, work with spatial data to solve problems.

Preface

The GIS Guide to Public Domain Data was written to provide GIS practitioners and instructors with the essential skills to find, acquire, format, and analyze public domain spatial data. The book provides context for the development of the numerous data portals we see today in terms of the organizations that collect and provide access to data, and the policies that govern data use. Although other documents have been published about public domain data, a critical evaluation of the various portals available and the merits of their data has been lacking. Each chapter in this book discusses the major issues associated with public domain spatial data by presenting both sides of the debate, helping you appreciate the complexity of these issues and their relevance in your every-day work with GIS. This information will help users in all areas of the GIS community become critical users of data—mindful of provenance, quality, and appropriate use.

Nowadays, it has never been easier to find and use spatial data. The plethora of data services and portals provide you, the GIS user, with more choices than ever before. However, it is not only easier to *use* data but also to *misuse* data. What are the relevant questions you should ask before you decide whether to use a particular set of data in your project? How do these questions affect your project's scope, management, work-flow, and ultimate success? How does your choice of data influence the results you will produce and, perhaps more importantly, the conclusions you will derive from those results? How do your choices affect the conclusions and decisions of your customers or clients who rely on your data?

This book brings together definitions of data and information, some of the vagaries of international copyright law, and the development of public access and public domain licensing. It provides information on the formats and themes of data that are available, how data are modeled in a spatial database, some of the issues surrounding data quality, data privacy, national and international standards, and metadata. This book reviews the free-versus-fee debate: Should public domain data be available at no cost or is cost justified in some cases? *The GIS Guide to Public Domain Data* also considers the national and international contexts for providing access to public domain data and looks specifically at the different approaches adopted in the United States and the United Kingdom. It looks at the phenomenon of volunteered geographic information, the advent of cloud computing, and the development of online GIS communities. These technological revolutions are illustrated in GIS practice with some timely examples of crowd-sourced, emergency mapping initiatives.

The online exercises that accompany each chapter enable you to put much of the book's theory into practice, using real data, relevant scenarios, and a methodical approach to solving problems with public domain spatial data. The examples include local, regional, and global-scale analyses. Some of the issues studied involve assessing natural hazards, locating a retail outlet, urban sprawl, land management, deforestation, and ecotourism. The exercises are designed to help you put spatial thinking and the geographic perspective into practice while considering the impact of the types of data, the accuracy and scale of those data, and the procedures run on those data on your final results.

About the exercises

The online exercises that accompany this book let you experience firsthand some of the many benefits, and occasional frustrations, involved in working with public domain data. They will direct you to long-established data portals to search for, download, and manipulate spatial data. The exercises address issues and solve real-world interdisciplinary problems using public domain data with GIS software (ArcGIS Online and ArcGIS for Desktop platforms).

We recommend that you first read each chapter to understand the concepts and then access the corresponding exercise to put theory into practice.

In exercise 1, you will analyze a set of global issues surrounding population, climate, and the environment. Specifically, you will examine world ecoregions that might be under the most population stress. You will locate and download data on aridity, coastlines, land cover, ecoregions, and population density, and then analyze the interaction among these phenomena: How does population density and climate change affect ecoregions and land cover? Which densely settled areas are in arid regions and could face water shortages? You will address these and other questions as you analyze data from the Food and Agriculture Organization of the United Nations.

In exercise 2, you have the opportunity to download and work with US Census Bureau demographic data and TIGER files to determine the best site to locate an Internet café in Orange County, California. Bear in mind as you work through the exercise that using a different set of demographic data, with different interpretation of campuses and neighborhoods, might have a significant influence on your choice of location for the Internet café.

Exercise 3 focuses on using public domain spatial data from several different government agencies to site a wildfire observation tower in the Loess Hills in eastern Nebraska. During the course of your analysis, you will examine and combine vector and raster data—land cover, elevation, hydrology, and transportation data—and from these data sources, you will create derived information products. In so doing, you are using GIS to make real decisions about everyday issues.

In exercise 4, you will use some of the spatial data from a local government agency and a university to assess and evaluate past and present flood events in Boulder, Colorado, as part of a 100-year and 500-year flood risk assessment exercise. In so doing, you are using GIS as part of a decision-making system that relies on accurate data inputs and also relies on your expertise as the data user and analyst.

Exercise 5 gives you the opportunity to use public domain spatial data from a variety of federal and state portals to assess hurricane risk on the Texas coastline. Through spatial analysis and the examination of combined vector and raster datasets created at different resolutions and scales, you will create derived information products using GIS analysis techniques. In so doing, you are using GIS to make real decisions about a timely and critical issue—natural hazards.

For exercise 6, you will use an international public domain spatial data portal to address a regional and national problem that is also a global issue—deforestation. You will assess the spatial pattern of land use and deforestation in Rondônia, a state in western Brazil, and combine data from the Brazilian government with Landsat imagery from an international data portal to assess the situation and offer recommendations.

Exercise 7 adopts a different format by allowing you to plan a small project for yourself. This will include deciding what data you need, where you will get those data, how you will use the data, and what type of information products will meet the requirements of your project. You will focus on producing a map for an ecotourism company in New Zealand.

In exercise 8, you will assess the utility of citizen science projects and, as you do so, think about issues of data quality and data access. You will also examine a popular citizen science initiative by studying invasive species, including zebra mussels and other species from the US National Atlas, and using an online GIS portal and in a desktop GIS environment. How do invasive species spread and change, and how are these patterns affected by land cover and land cover diversity?

In exercise 9, you will use cloud-based GIS services in a variety of ways—accessing and creating online mapping mashups, using online mapping tools, and streaming data from the cloud via ArcGIS Online. You will use these skills and public domain spatial data, together with some spatial analysis techniques, to assess three major hazards of 2010: The Gulf of Mexico oil spill, the eruption of the Eyjafjallajökull volcano in Iceland, and the earthquake in Haiti.

Finally, for exercise 10, you will have the opportunity to select the most suitable locations for expanding tea cultivation in Kenya using GIS and spatial analysis. You will use public domain spatial data from the World Resources Institute for your analysis, with data on elevation, precipitation, land use, wetlands, rivers and lakes, urban areas, protected areas, roads, and zoning.

As the Internet is a continually changing landscape, with websites appearing and disappearing with relative ease, we have chosen several data portals supported by major organizations to provide the raw data for these activities. Even so, there's no guarantee that when you work through some of the exercises, the specific data portal you require will be available. To avoid delays and enable you to complete the exercises in a timely manner, we have provided backup copies of the data on ArcGIS Online. To access, visit ArcGIS Online at: http://www.arcgis.com. In the Search window, select Search for Groups and then enter **GIS Guide to Public Domain Data**. Select the group that appears; this will take you to the list of exercises. Select the exercise you wish to work on, and click the Open arrow underneath the thumbnail map to download the data. Save the data to an appropriate folder on your computer. Unzip this file using a file extraction utility such as WinZip or 7-Zip. You will see a number of data files and an ArcMap document (.mxd). Once you have downloaded and unzipped the data, you can skip ahead in the exercise to the point where you begin working with the data.

SPATIAL DATA AND THE PUBLIC DOMAIN

Introduction

Spatial data provide the foundation for making decisions with geographic information systems (GIS). Without real-world data, GIS analysis is reduced to theory and conjecture. Wilson (2001) may have summed it up best: "After more than thirty years, we're still confronted by the same major challenge that GIS professionals have always faced: You must have good data. And good data are expensive and difficult to create" (p. 54).

Today, spatial data are available in more formats, obtainable by more methods, produced by more agencies and organizations, at more scales, in more themes, and at a higher resolution than ever before. Much of the data are in the public domain and literally at the fingertips of nearly every GIS user, ready to be streamed or downloaded and used in the decision-making process. Why then is there a need for this book? Despite the acknowledged importance of having access to good data, the increasing availability of public domain spatial data makes it tempting to plunge right in, download, and start using data within a GIS environment without knowing a great deal about the data:

How do you know if the data are suitable for your project?
How do you know if what you are planning to do with the data is appropriate?

Where did the data come from, who collected them, how were they collected, at what scale were they collected, and are they accurate enough for your project?

It is also important to understand that not all publicly available spatial data are in the public domain. Just because a dataset is available to download from a website, that does not guarantee it is free from copyright and licensing restrictions, or possibly subject to a fee.

Failure to appreciate both the benefits and limitations of using public domain spatial data sources will inevitably lead to, at best, inappropriate decisions with unforeseen economic, legal, or environmental consequences. At worst, lives and property could be at risk as a result of flawed analysis.

Equally as important as gaining knowledge about data and practicing skills in using them is a thorough understanding of the related issues and policies at the local, regional, and global levels. The hardware, software, operating systems, the Internet, and tools to manipulate data are continually changing, but changing just as rapidly are the environments in which all these components operate. This includes the organizations that produce and use spatial data, political systems, national and international standards, spatial data policies and laws, and finally user expectations. How can you, the GIS data user, keep track of these changes to fully appreciate the implications for your projects?

This book has five objectives:

1. To help you understand the sources, scales, accuracy, benefits, and limitations of spatial data.
2. To improve your awareness of the organizations and initiatives that produce spatial data, including public agencies (local, state, federal, tribal, and international), private companies, universities, nonprofit organizations, and multi-organization collaborative initiatives.
3. To enable you to become familiar with the history, current issues, and policies—both domestic and international—that affect the production of and access to public domain spatial data.
4. Through a series of associated exercises, to help you develop a working knowledge of, and practical skills in finding, accessing, downloading, and formatting public domain raster and vector data.
5. To augment your spatial analysis skills to effectively address issues and solve real-world, interdisciplinary problems by using public domain data with GIS software (Esri's ArcGIS, both the online and the desktop platforms).

This book provides some practical guidance that will help you make more informed decisions about the choice of data sources for your projects. It also provides a reference for the types, formats, and variety of spatial data that are available and some background information on the often complex organizations that collect and manage those data. The accompanying exercises include siting a business, assessing flood potential, evaluating land use changes, and analyzing the historical impact of hurricanes.

This book also aims to foster critical and spatial thinking. Critical thinking is a combination of skills and techniques that start with asking questions, acquiring and reviewing evidence, and lead to deducing sound and substantiated conclusions. In the context of GIS, critical thinkers need to be skeptical of data and the problem-solving strategies that are employed. They need to resist the temptation to accept the easiest-to-use datasets or the first ones they find, and instead identify only those most appropriate to solve their particular problem. Critical thinking also requires understanding motives and biases, both in ourselves and in the sources of information we use.

Spatial thinking means adopting a geographic perspective that considers scale, temporal and place-based relationships, patterns, linkages, and trends. Geography is not a body of content to be memorized. Facts and figures are important, but all successful GIS users, regardless of their fields of interest, need to employ geographic skills. These skills are honed through spatial analysis and the tools we used to interrogate the data such as summarizing tabular data, computing mean centers, intersecting, and buffering. At each stage in that analysis, step back from the functions and ask why each procedure is being performed and what effect each has on the data and on your final results. This ensures the geographic perspective—a way of looking at and understanding the world—is embedded throughout the process. The chapters and the exercises in this book are designed for you to develop your critical and spatial thinking skills.

The target audience for this book is anyone who is working with, or may be considering, public domain spatial data as part of GIS-based project. The level of GIS skill required to work through the exercises is intermediate, a working familiarity with ArcGIS for Desktop and perhaps some experience in file handling and data management. The exercises are not solely concerned with improving your analytical GIS skills. They are designed to encourage you to think about what types of data you need to solve a particular problem, the steps required to get the data into the correct format for your GIS, the quality of the data, and what other datasets you may need to address the problem. Other issues raised in the chapters are applied to real-world problems.

The text is written for university-level and other adult learners, including students, professors and researchers in universities, GIS practitioners in government agencies, nonprofit organizations, and private enterprise. We want you to develop the necessary skills for finding, understanding, downloading, using, and ultimately solving problems with public domain spatial data. Throughout the course of the book, we will review the policies, issues, history, and current developments in the provision and use of spatial data in GIS-based analysis. Although many websites providing access to or information about spatial data are referenced in the book, these are not the only resources available. By necessity, we have chosen a limited number of examples to illustrate particular points.

Each chapter of the book focuses on a major theme, such as data types and formats, spatial data infrastructures, and planning GIS projects. Each chapter also includes a discussion of issues associated with public domain spatial data, such as whether organizations should charge fees or offer their data for free. Also included are additional references for those who wish to investigate issues on their own. The exercises associated with each chapter invite you to apply what you have learned to a real problem using public domain spatial data and ArcGIS software.

Given the range of topics and issues to be covered, perhaps the logical place to start is by introducing the concepts of data and information, including definitions of data, spatial data, and finally public domain spatial data.

Data versus information

Although many people use *data* and *information* interchangeably, these two terms actually convey very different concepts. Data are facts or figures which have been gathered systematically for one or more specific purposes. Data can exist in the form of linguistic expressions such as name or age, symbolic expressions such as traffic signs, mathematical expressions such as $e=mc^2$, or signals such as electromagnetic waves. Information, on the other hand, is data which have been processed into a form that is meaningful to a recipient and has perceived value in current or prospective decision making.

Although data are essential precursors for information, not all data make useful information. For example, data not properly collected, organized, or documented may be a hindrance rather than an asset to someone seeking to extract information. How useful data are also varies with each individual and application; data that generate useful information for one person's application may not be useful for another application. Information is only useful to its recipients when it is *relevant* for its intended purposes and with the appropriate level of detail. Information is also only useful when it is reliable, accurate, and verifiable (by independent means), and only when it is up-to-date, timely, and complete (in terms of its attributes, and its spatial and temporal coverage). Information must be intelligible, that is, comprehensible by its recipients, consistent with other sources of information, convenient (easy to handle), and adequately protected. As such, much of the public domain spatial data we will be using in this book are really information, because the data have already been processed into a form that is useful to GIS users. However, gaps in the documentation about the information, or indeed the information itself, may prevent it from being used to its full potential, as we shall see later on. Whether that information should be used or not, and the pros and cons of doing so, is one of the major themes of this book.

The primary function of an information system is to change data into information using one of the following four processes:

Conversion: The conversion process transforms data from one format to another, from one unit of measurement to another, and/or from one feature classification to another.

Organization: Organization processes arrange data according to database management rules and procedures so that they can be stored and accessed effectively.

Structuring: The structuring process formats or reformats data so that they can be acceptable to a particular software application or information system.

Modeling: Modeling processes include visualization and analysis that improve the user's knowledge base through the ability to emulate real-world processes and document workflows. This allows procedures and techniques to be replicated and the outcome predicted and assists the decision-making process.

The concepts of organization and structure are crucial to the functioning of information systems: Without organization and structure, it would be impossible to turn data into information.

Spatial data

The term spatial data is often used synonymously with geographic data and geospatial data. Spatial data are considered by some to be wider in scope than geographic data by including anything about space that doesn't necessarily have geographic coordinates, such as an MRI (magnetic resonance imaging) scan or CAD (computer-aided design) drawings of a building. For the purpose of this book, the terms are used interchangeably. The book focuses on data that are pertinent to features above, on, or below the surface of the earth—in other words, those having real-world coordinates. Such data are collected, manipulated, and used to make decisions in association with some aspect of geography, such as the context, distribution, or spatial relationships at a given location. Consider the following two seemingly diverse issues: crime in a neighborhood and soil erosion in a county. Each occurs at different scales and at different rates. Each affects, and is affected by, people in different ways, yet manifests itself as geographic patterns and phenomena which may be represented by a single layer or multiple layers of spatial data.

Spatial data are geographically referenced; they are identified and located by coordinates. These coordinates, in turn, may be part of a global, national, or regional referencing system but all include a point of origin, axes, and measurement units. Spatial data include information about where they are located and their extent. Spatial data are stored as points, lines, areas, or pixels, and include a symbol, or set of symbols to represent the data. Most spatial data exist within a topological framework, which determines how data items in the same layer or in different layers are spatially related to one another. These elements can be thought of collectively as the G part of GIS. The elements represent real-world features or phenomena, such as ocean currents, soil types, or fault lines. Spatial data also contain a descriptive element that informs the users what the data represent. The descriptive element can be thought of as the *I* part of GIS, or *information*. This could be the name of the ocean current, the pH of the soil type, or the strike and dip of the fault line. The relationship between individual data items reflects the *S* or *systems* part of GIS. If specific soils are queried or selected on the map, the resulting selection is reflected in the table containing the attributes. Conversely, if the soils types are selected in the attribute table, the corresponding feature selection is displayed on the map. The graphical element is commonly referred to as spatial data, while the descriptive element is commonly referred to as nonspatial or attribute data. Taken together, these elements may be used to represent spatial phenomena, issues, and objects as mappable items.

Spatial information is obtained by processing geographic data, the aim of which is to improve the user's knowledge and understanding of the geography of features and resources. This will help promote a better understanding of the consequences of human activities associated with those features and resources by developing spatial intelligence for problem solving and decision making.

Some will argue that spatial data are special data, but for others that designation simply reinforces the notion that spatial data are difficult to use and remain the preserve of the specialist and technically minded. Capturing, managing, and maintaining spatial data does involve some specialized skills, but most people who work with spatial data are not so concerned about how the data are captured, stored, and updated. Rather, they just care that the data are available when they need it, that the source and any use restrictions are clearly documented, the data are easy to use, and help complete their projects.

Using spatial data and information to solve problems is nothing new. Some of the earliest known maps date back to prehistoric times and can be found in the cave paintings and engravings of the earliest civilizations. Although its interpretation as a map is debated by some, the markings on a fragment of mammoth tusk, excavated in Mezhirichi, Ukraine, in 1966, are thought to represent some dwellings and the location of fishing nets on the bank of a river (Harley and Woodward 1987). Examples from ancient Egypt include the Turin Papyrus Map, believed to date from 1160 BC, which was prepared to assist in recording the location of rock suitable for creating sculptures. The map was annotated with the location of gold deposits, the destinations of some of the wadi routes, and distances between quarries and mines, among other things.

One of the best examples from Roman times is the *Forma Urbis Romae*, a highly detailed ground plan of every building and monument in Rome. The plan was incised onto marble slabs that hung on a wall in the Templum Pacis in Rome. The exact purpose of the plan is unclear, but given its size and location, it is thought by some to be a decorative representation of the more utilitarian cadastral mapping of the time that may have had some role in urban planning. Also from Roman times, the Peutinger Map or *Tabula Peutingeriana*, a copy of a Roman map made in the thirteenth century, shows an elongated schematic of the road network in the Roman Empire. The map depicts Roman settlements and the roads connecting them, rivers, mountains, forests, and seas. In addition, the distances between the settlements are recorded in a variety of measurements.

During the Age of the Enlightenment, the use of maps greatly expanded as interest arose in faraway places. Maps stirred imaginations and inspired explorations of the unknown. Yet mapmaking was reserved for the select few who had access to the data from surveyors and explorers and also possessed the tools and the necessary skills to bring those data to life. Having access to accurate mapped information bestowed a great deal of power on those who were able to exploit it for their own gains.

No longer the preserve of the intelligentsia, the production of, and access to, spatial data is evolving in the twenty-first century at ever-increasing speeds. By 2010, more than one million servers existed around the world; every day, Google alone processed one billion search requests and 20 petabytes (1000 terabytes or 1,000,000 gigabytes) of user-generated data. As much of this data are in the form of custom-made maps, it is probably safe to say that more maps are now made each month than the sum total of maps made in all recorded human history. The amount of spatial data available to GIS practitioners is greater now than at any time in the fifty-year history of GIS.

In the early days of GIS, when comparatively little digital data were available, a standard piece of hardware next to nearly every GIS workstation was the digitizing tablet; the first task in most projects was to collect data. While GIS users today still generate some of their own data, the amount of data amassed from decades of investment by government agencies, academia, industry, and nonprofit organizations means that many users can simply tap into the vast reserves of data created by others for their own projects.

Not only has the amount of available data increased, but also the variety of spatial data has greatly improved. During the 1980s, geographic base files/dual independent map encoding (GBF/DIME) files from the US Census Bureau along with Digital Line Graphs and Landsat satellite imagery from the US Geological Survey (USGS) comprised the bulk of spatial data available in the United States. Federal agencies were responsible for the majority of data production, as only they had the necessary staff and technical resources required to convert paper maps to digital data and create new digital data from satellite imagery. Nowadays, spatial data are available for hundreds of themes and phenomena, literally from A to Z—agriculture to zebra mussels. In addition, the number of data formats has also expanded. Data are available in raster form, in vector form, as static documents, and through real-time dynamic services. At the same time, the ways to access and obtain data have never been greater. Data can be downloaded as files and stored on a local computer, data can be streamed from the web without having to store a copy locally, or data may be ordered on physical media.

The number of organizations providing spatial data increased steadily through the first decade of the 2000s. Higher educational institutions, nonprofit organizations, private companies, and local and regional governments joined ranks with the federal agencies as the top data providers. As the decade came to a close, individual GIS users contributed to national and international data collection initiatives on an unprecedented scale thanks to easy-to-use application programming interfaces (APIs), broadband access, and the networking and hardware infrastructure necessary to support this emerging online community. Known as citizen science or crowdsourcing, GIS users can now regularly provide their own volunteered geographic information. The new paradigm of citizen science brings renewed attention to familiar data issues:

Can user-generated data be trusted?

How can anyone using the data understand the contexts, scales, accuracy, and situations where these data should be used or, perhaps more importantly, should not be used?

What is the role of traditional data providers in this new era?

Despite the tremendous expansion in the amount, variety, formats, means of distribution, and number of data providers, challenges still remain today for those wishing to work with spatial data. Perhaps one of the greatest challenges now is locating the most appropriate spatial data for a particular project. It has become very easy to establish a data portal, resulting in the ever-increasing number and variety of portals we see today. However, not all portals are created equally. Some are well documented, while others are not. Some work only with certain web browsers, while others require web browser settings to be adjusted or browser plug-ins to be installed. Some cover multiple regions, such as the SERVIR Regional Visualization and Monitoring System website where some sections are populated with more data (SERVIR Mesoamerica), while others are less populated (SERVIR Africa). There is still no comprehensive and all-inclusive one-stop shop for finding spatial data.

Obtaining spatial data

There are several ways to obtain spatial data. You could purchase data from an organization or individual that has already collected (or will collect) them for you or download data for free. Alternatively, you could create the data yourself: gather the relevant attribute data along with spatial coordinates from a GPS-enabled device, geocode points from a set of address data, scan paper maps or imagery, or gather data via other means. Regardless of which method you choose, there are associated costs and benefits in terms of finances, time, data quality, copyright, and other issues, as we shall examine later.

This book focuses on downloading or streaming spatial data and, in particular, data that are available for free with subsequent usage not subject to copyright or licensing constraints. In other words, public domain data. To appreciate what public domain data can offer GIS analysts and end users, you need to understand what public domain means and what can you do with the data from those sources. It is also important to understand the impact public data have had on the use and application of GIS in the past and the impact they will have in the future.

The public domain

Numerous definitions for the term *public domain* exist. Although common themes in most of these definitions include the lack of copyright, the freedom from intellectual property rights (IPR), and no requirement to pay royalties, there is no universally accepted definition of public domain.

The US Copyright Office (2011) states: "The public domain is not a place. A work of authorship is in the 'public domain' if it is no longer under copyright protection or if it failed to meet the requirements for copyright protection. Works in the public domain may be used freely without the permission of the former copyright owner."

In the United Kingdom, the Intellectual Property Office (2011) defines public domain as "The body of works not or no longer protected by Intellectual Property rights which are available for the public to use without seeking permission or paying royalties."

Creative Commons (2011), a nonprofit organization established in 2001 to make "it easier for people to share and build upon the work of others, consistent with the rules of copyright," defines the public domain as "When a work is in the public domain, it is free for use by anyone for any purpose without restriction under copyright law. Public domain is the purest form of open/free, since no one owns or controls the material in any way."

The United Nations Educational, Scientific, and Cultural Organization (UNESCO) (2011) defines the term as "Public domain information is publicly accessible information, the use of which does not infringe any legal right, or breach any other communal right (such as indigenous rights) or any obligation of confidentiality."

The Linux Information Project (LINFO) (2011) defines it as "Public domain refers to the total absence of copyright protection for a creative work (such as a book, painting, photograph, movie, poem, article, piece of music, product design, or computer program)."

Wikipedia (2011) refers to public domain as "Works are in the public domain if they are not covered by intellectual property rights at all, if the intellectual property rights have expired, and/or if the intellectual property rights are forfeited."

As a general rule, you should think of public domain spatial data as publicly accessible information about a spatial theme or phenomenon, the use of which does not infringe the legal rights of an individual or organization. Public domain spatial data encompasses all works or objects that may be used by anyone without any authorization.

Not only does public domain mean different things to different data users and data providers, the definition also varies from country to country. In fact, the concept of public domain simply does not exist in countries with no copyright laws, where the country owns all property including intellectual property, or where the country designates large amounts of information as state secrets or withheld in the interests of national security. As a further complication, copyright law is sometimes amended, for example when the US government extended copyright twenty years beyond the original terms stated by the Copyright Act of 1976. Such changes are controversial and subject to challenge, as in the case of *Eldred v. Ashcroft*, which sought to overturn the US government's extension of the copyright period. As a result of the extension, a number of works were prevented from entering the public domain in 1998 and following years.

One area of difficulty applying national copyright laws to spatial data is that spatial data by their very nature often deal with phenomena that have no respect for human-made boundaries and borders. Which country's copyright laws apply in such international situations? Another difficulty in applying copyright to spatial data arises when datasets are derived from multiple sources. It is often difficult to trace which organization or individual originally gathered specific data, or to separate and identify discrete data items, and therefore to determine if those data should be copyrighted. Some spatial data also combine elements from old and new sources. Features such as contour lines from topographic maps may have been created in analog form on paper, vellum, or even copper plates decades or a century ago but may appear together now with new satellite imagery. What effect does the date of creation have on whether a work is considered to be in the public domain?

Although there are many similarities as to what the public domain means to different national bodies, there is a lack of consensus at the international level. If a work is made available in the public domain in one country, what effect does that have on its copyright status in another country? What you can do with a certain dataset in one country may not apply in another country, even though you are still working with the same data? The US Copyright Office posts the following advisory on its website: "Even if you conclude that a work is in the public domain in the United States, this does not necessarily mean that you are free to use it in other countries. Every nation has its own laws governing the length and scope of copyright protection, and these are applicable to uses of the work within that nation's borders. Thus, the expiration or loss of copyright protection in the United States may still leave the work fully protected against unauthorized use in other countries."

FOR FURTHER READING

If you are interested in learning more about copyright and public domain in the United States, Cornell University (2011) provides a useful report, updated annually, identifying "works" that are in the public domain.

With respect to individual data providers, the terms and conditions for using those data are also varied. Companies such as Yahoo! and MapQuest have been providing web-based map services for a number of years. Although these data are generally available to the public, the data they provide access to are made available under proprietary licensing arrangements that impose certain restrictions on how the data may be used. Remember that there is a clear difference between what is *publicly accessible* versus what is in the *public domain*. Just because a dataset is downloadable does not place it in the public domain. Many photographs, for example, are available online today but are copyrighted, and some have a copyright symbol or watermark on the image itself. Spatial data are less likely to carry watermark or copyright symbols, but even if these are lacking, do not assume the data are in the public domain.

For Yahoo! Maps, the terms of use include

"You agree to use the Data together with Yahoo! Maps solely for personal, non-commercial purposes for which you were licensed, and not for service bureau, time-sharing or other similar purposes. Accordingly, but subject to the restrictions set forth in the following paragraphs, you may copy this Data only as necessary for your personal use to (i) view it on your screen, (ii) print it, (iii) transfer a copy to a personal electronic device such as a personal digital assistant; (iv) save it, and (v) transfer a copy, in html form only, to a third party provided that you do not remove any copyright notices that appear and do not modify the Data in any way. You agree not to otherwise reproduce, copy, modify, decompile, disassemble or reverse engineer any portion of this Data, and may not otherwise transfer or distribute it in any form, for any purpose, except to the extent permitted by mandatory laws."

Similarly, for MapQuest, the terms of the license are as follows:

"MapQuest grants you a nonexclusive, non-transferable license to view and print the Materials solely for your own personal non-commercial use. You may share a map or directions with another individual for that individual's personal non-commercial use using the email option on the webpage. You may not commercially exploit the Materials or the underlying data, including without limitation, you may not create derivative works of the Materials, use any data mining, robots, or similar data gathering and extraction tools on the Materials, frame any portion of the Materials, or reprint, copy, modify, translate, port, publish, sublicense, assign, transfer, sell, or otherwise distribute the Materials without the prior written consent of MapQuest. You shall not derive or attempt to derive the source code or structure of all or any portion of the Materials by reverse engineering, disassembly, decompilation or any other means. You shall not use the Materials to operate a service bureau or for any other use involving the processing of data of others."

As such, Yahoo! Maps and MapQuest Maps are not in the public domain.

Similarly, for those using Google Maps and Google Earth, the following restrictions on use terms are imposed:

"Unless you have received prior written authorization from Google (or, as applicable, from the provider of particular Content), you must not:

A. access or use the Products or any Content through any technology or means other than those provided in the Products, or through other explicitly authorized means Google may designate (such as through the Google Maps/Google Earth APIs);

B. copy, translate, modify, or make derivative works of the Content or any part thereof;

C. redistribute, sublicense, rent, publish, sell, assign, lease, market, transfer, or otherwise make the Products or Content available to third parties;

D. reverse engineer, decompile or otherwise attempt to extract the source code of the Service or any part thereof, unless this is expressly permitted or required by applicable law;

E. use the Products in a manner that gives you or any other person access to mass downloads or bulk feeds of any Content, including but not limited to numerical latitude or longitude coordinates, imagery, and visible map data;

F. delete, obscure, or in any manner alter any warning, notice (including but not limited to any copyright or other proprietary rights notice), or link that appears in the Products or the Content; or

G. use the Service or Content with any products, systems, or applications for or in connection with (i) real time navigation or route guidance, including but not limited to turn-by-turn route guidance that is synchronized to the position of a user's sensor-enabled device; or (ii) any systems or functions for automatic or autonomous control of vehicle behavior."

Google Maps and Google Earth are frequently used for thousands of everyday tasks, even providing context mapping for the display of other data that may be in the public domain. However, as with Yahoo! Maps and MapQuest Maps, Google map products are not free from restrictions on use and are not in the public domain.

You should also be aware of situations where a public domain spatial data source becomes part of another data source. OneGeology, an international geological spatial data online resource, was established in 2007 to "create dynamic geological map data of the world, available to everyone via the web." Most of the 116 contributors to the site are national geological mapping agencies, such as the USGS, the British Geological Survey, the Geological Survey of Pakistan, and the Geological Survey of Japan. The terms and conditions that govern access to data hosted on the site warn against copyright infringement if the data are used for commercial purposes. As such, this compilation of data from OneGeology is not in the public domain. However, some of the individual datasets that make up OneGeology may indeed be in the public domain. The OneGeology (2011) use agreement states "Where you intend to use the material commercially (e.g. in book, to sell as a map extract, etc.) you will need to pay the appropriate copyright holder which will be the owner of the geological information from the country concerned." Although hosted by an international consortium and publicly available to anyone with Internet access, the data remain subject to the individual copyright restrictions of each national mapping and data collection agency. Regardless of them being a component of OneGeology, the USGS data are in the public domain because the data are produced by a US federal agency under section 105 of the Copyright Law of the United States. You should be aware of the distinction in copyright terms between compilations and the items that make up those compilations which may have different use constraints.

One example of an online spatial data resource in the public domain is the Natural Earth website. Natural Earth data were created by cartographers who collaborated in a project supported by the North American Cartographic Information Society (NACIS). Natural Earth provides global small-scale raster and vector maps, based on three themes:

- Cultural, including countries, urban areas and administrative boundaries.
- Physical, including land, coastline, rivers and lakes.
- Raster, including shaded relief, bathymetry, and ocean.

The maps are at three different scales: 1:10,000,000; 1:50,000,000; and 1:110,000,000. The terms of use are as follows:

"All versions of Natural Earth raster + vector map data found on this website are in the public domain. You may use the maps in any manner, including modifying the content and design, electronic dissemination, and offset printing. The primary authors . . . and all other contributors renounce all financial claim to the maps and invite you to use them for personal, educational, and commercial purposes."

Another example of an online spatial data resource is the North American Environmental Atlas, compiled by the Commission for Environmental Cooperation (CEC). The CEC was set up around the time the North American Free Trade Agreement (NAFTA) took effect, and includes the same countries that participate in NAFTA: Canada, the United States of America, and Mexico. The commission was established by a multinational agreement to address regional environmental concerns, help prevent potential conflicts between trade and the environment, and to promote the effective enforcement of environmental law. The commission created the 1:10,000,000-scale North American Environmental Atlas, a public domain dataset with an accompanying data viewer. Although data viewers provide a quick and easy way to inspect the data, they are of limited use for most GIS users as many do not support any analytical capabilities. However, the North American Environmental Atlas goes one step further in providing data files that can be downloaded and used by the GIS community. As the atlas contains data from a variety of sources, the terms of use restrictions provided on the CEC website accounts for these sources as follows:

"The creator of this data is outlined in the metadata and at the bottom left corner of each MXD or GeoPDF that is made available through the Commission for Environmental Cooperation (CEC). If the author of this data is the CEC: The user is permitted to use this data for non-commercial, non-sublicensable purposes. The user is allowed non-exclusive rights to represent this data as desired. If the author of the data is noted otherwise: Please refer to the license information of this individual dataset, or contact that organization or individual if you are unsure of your rights regarding redistribution, use, or license information of this particular dataset."

The commission also provides disclaimers on the quality of the data in the atlas:

"The CEC does not ensure that this is the most updated representation of this information, although updates will be conducted on an irregular basis. We make every effort to provide and maintain accurate, complete, usable, and timely information on our Web pages. However, some CEC data and information accessed through these pages may, of necessity, be preliminary in nature. These data and information are provided with the understanding that they are not guaranteed to be complete. Users are cautioned to consider carefully the provisional nature of these data and information before using them for decisions that concern personal or public safety or the conduct of business that involves substantial monetary or operational consequences. Conclusions drawn from, or actions undertaken on the basis of, such data and information are the sole responsibility of the user."

The commission's license arrangements define how the data should be cited, documents the limitations of the data, and indicates who may use the data and for what purpose. The intention is for the user to not assume that every area feature is up-to-date. The data are in the public domain and are to be used as such. In common with many similar organizations, there are restrictions on the use of certain datasets for commercial purposes, just as there are restrictions on certain software packages for educational versus commercial use.

The few examples described here demonstrate that although a great deal of spatial data are publicly available, those data are not always in the public domain. Anyone wishing to make use of a publicly available spatial data resource should always *read the label*. That said, the amount of available public domain spatial data has expanded rapidly in recent years and has become an integral part of the evolving GIS community. However, no discussion of the public domain would be complete without also considering copyright and licensing, a subject we touched on briefly earlier but which requires some elaboration.

FOR FURTHER READING

See James Boyle's *The Public Domain: Enclosing the Commons of the Mind* (Caravan Books 2008). Boyle is a professor at Duke University School of Law and founder of the Center for the Study of the Public Domain. This book is available online under a Creative Commons Attribution-Noncommercial-Share Alike License.

Copyright

What is copyright? In its simplest form, copyright is a form of protection provided by laws to authors who create *original works*. These works can be art, music, drama, literature, and—most important to GIS users—data. Copyright laws grant the authors of these works certain exclusive rights to do such things as use the works, reproduce the works, prepare derivative works, and distribute copies. Copyright laws also give authors the right to authorize who may copy and use the works they created, and when they may do so. Registration of the copyright brings the author additional legal rights (Patterson and Lindberg 1991).

Section 105 of the Copyright Law of the United States, titled "Subject matter of copyright: United States Government works," has important implications for GIS users. It states that "Copyright protection under this title is not available for any work of the United States Government." As the works of the US government, including spatial data, cannot be copyrighted, the spatial data created by US federal agencies have been available to and used by the GIS community. This particular piece of legislation resulted in an entire industry built upon government spatial data, as we shall see in this book, through its impact on privacy and fee structures. The amount of spatial data generated by federal agencies in the United States dwarfed the amount of data from federal agencies in other countries where, with a few exceptions, data are subject to copyright and licensing restrictions. This situation is, however, changing and will be discussed in greater detail in chapter 4.

Another aspect we will explore is that section 105 only applies to US federal government works; state and local spatial data can be copyrighted and licensed. A landmark case in the application of copyright in the United States was *Feist Publications Inc. v. Rural Telephone Service Co.* The case centered on the licensing of telephone directories and an alleged copyright infringement. The court decided that the listings contained in the directory were not subject to copyright so no infringement had taken place. By implication, facts alone were not copyrightable; a "minimal degree" of creativity must be involved to make something copyrightable. As such, any private company that enhances, or adds value to, US government spatial data may indeed copyright those data.

In 1989, the United States became a signatory to the Berne Convention for the Protection of Artistic and Literary Works, the worldwide agreement on copyright protection dating back to 1886. The Berne Convention provisions, which are codified by every adopting country, set forth the principle that once a work is deemed completed by its author, the work is automatically copyrighted. This principle was established to overcome loss of copyright protection due to the filing and registration technicalities that some early copyright laws (such as the 1909 US Copyright Act) invoked. As a general rule, everything is copyrighted unless there is explicit language to the contrary. That explicit language is an affirmative statement that the author is making the work available free of copyright and in the public domain. Works come into the public domain for two major reasons; the duration or term of copyright lapses or expires or the author specifically places the work in the public domain.

It is illegal for anyone to violate any of the rights provided by the copyright law to the owner of copyright. However, the rights are not unlimited. For example, the US Copyright Act of 1976 established some exemptions, including one that affects many GIS data users: the doctrine of *fair use* (section 107). Fair use is a difficult concept to define, but four factors in US law help determine if a particular use is fair. These factors include the purpose and character of the proposed use (such as commercial or educational), the nature of the copyrighted work, the amount used compared to the copyrighted work as a whole, and the effect of the use upon the potential market or value of the work. Another example of a limitation of copyright law occurs when royalties are paid on some works and conditions are agreed to by the purchaser, the purchaser may use the copyrighted work under those conditions.

Creative Commons licensing and the public domain

What does a public domain license really mean and how it is administered? As we discussed earlier, the Creative Commons organization was specifically established to promote and facilitate access to what they refer to as "creativity (cultural, educational, and scientific content) in 'the commons.'" To support this initiative, Creative Commons provides a set of licensing options for individuals and organizations alike to grant the required level of access to creative work produced by those individuals or organizations. The level of access may range from full copyright protection and all rights reserved to certified as public domain and no rights reserved. A summary of the main licenses and conditions, starting with the least restrictive and ending with the most restrictive, is provided in the following:

Attribution (CC BY): Others may distribute, remix, tweak, build on the work, and use for commercial purposes as long as originator is given credit. http://creativecommons.org/licenses/by/3.0/

Attribution-ShareAlike (CC BY-SA): Others may distribute, remix, tweak, build on the work, and use for commercial purposes as long as originator is given credit and all new creations are licensed under identical terms. http://creativecommons.org/licenses/by-sa/3.0/

Attribution-NoDerivs (CC BY-ND): Allows for redistribution, commercial and noncommercial, as long as it is unchanged and in whole, and originator is given credit. http://creativecommons.org/licenses/by-nd/3.0/

Attribution-NonCommercial (CC BY-NC): Others remix, tweak, and build upon the work noncommercially, and although their new works must also acknowledge the originator and be noncommercial, they don't have to license their derivative works on the same terms. http://creativecommons.org/licenses/by-nc/3.0/

Attribution-NonCommercial-ShareAlike (CC BY-NC-SA): Others may remix, tweak, and build upon the work noncommercially, as long as they credit the originator and license their new creations under the identical terms. Others can download and redistribute the work, translate, make remixes, and produce other work based on the work. All new work will carry the same license, so any derivatives will also be noncommercial. http://creativecommons.org/licenses/by-nc-sa/3.0/

Attribution-NonCommercial-NoDerivs (CC BY-NC-ND): Others may download the work and share with others as long as they mention and link back to the originator. The work cannot be altered in any way or used commercially. http://creativecommons.org/licenses/by-nc-nd/3.0/. (Source: http://creativecommons.org.)

For those individuals and organizations that may wish to make their work available with no conditions, or to have it certified as in the public domain, Creative Commons introduced the concept of CC0: "CC0 which in effect removes all copyright from a piece of work or data with a No Rights Reserved license." CC0 allows content originators and owners to waive copyright interests in their work and place them in the public domain worldwide.

Creative Commons public domain license logo.
Courtesy of Creative Commons.

Creative Commons has also recognized the need for global consensus in the interpretation and application of these licensing arrangements and is working to translate and adapt these agreements to international copyright jurisdictions.

FOR FURTHER READING

For a complete list of the countries that now have their licensing arrangements integrated into the Creative Commons licensing process, see the CC Affiliate Network on the Creative Commons website.

To assist data providers in creating an affirmative statement that the author is placing the work in the public domain, Creative Commons offers a Copyright-Only Dedication (based on US law) or Public Domain Certification. The wording reads as follows:

"The person or persons who have associated work with this document (the 'Dedicator' or 'Certifier') hereby either (a) certifies that, to the best of his knowledge, the work of authorship identified is in the public domain of the country from which the work is published, or (b) hereby dedicates whatever copyright the dedicators holds in the work of authorship identified below (the 'Work') to the public domain. A certifier, moreover, dedicates any copyright interest he may have in the associated work, and for these purposes, is described as a 'dedicator' below. A certifier has taken reasonable steps to verify the copyright status of this work. Certifier recognizes that his good faith efforts may not shield him from liability if in fact the work certified is not in the public domain. Dedicator makes this dedication for the benefit of the public at large and to the detriment of the Dedicator's heirs and successors. Dedicator intends this dedication to be an overt act of relinquishment in perpetuity of all present and future rights under copyright law, whether vested or contingent, in the Work. Dedicator understands that such relinquishment of all rights includes the relinquishment of all rights to enforce (by lawsuit or otherwise) those copyrights in the Work. Dedicator recognizes that, once placed in the public domain, the Work may be freely reproduced, distributed, transmitted, used, modified, built upon, or otherwise exploited by anyone for any purpose, commercial or non-commercial, and in any way, including by methods that have not yet been invented or conceived."

Two examples of publicly available, but not public domain, spatial data sources provided under a Creative Commons licensing arrangement are OpenStreetMap (OSM) and European Environment Agency (EEA). OSM is a collaborative project that aims to develop and provide a free, editable, map of the world. XML exports of the data are available for use under Creative Commons Attribution-Share Alike 2.0 license. EEA collates and produces integrated environmental datasets for the European Union. Data are available to download from the EEA website under a Creative Commons Attribution license. We will return to the OSM project in chapter 8 when we cover crowdsourced data.

The Creative Commons approach to licensing is not without disadvantage; it was never developed with spatial data and databases in mind, nor with the attendant issues of combined data stores and generating derived data. Another valid approach may be the Open Database License proposed by the OSM community, which would see the introduction of an Attribution Share Alike (CC BY-SA) license for databases. Although the Creative Commons approach is perhaps better suited to literary works rather than data, and Creative Commons does not recommend its license model for software, at present the Creative Commons approach appears to be emerging as the de facto licensing arrangement for spatial data.

The web and public domain data portals

Since its inception in the 1960s, GIS, like other aspects of information technology in the digital age, has undergone a rapid evolution. Over the same time span, data have also changed—their formats, resolution, themes and types, how they are stored, and how they are served. The main platform providing access to data—the web—is also changing rapidly, almost daily. Throughout the course of this book, you will be referred to websites and some of the URLs will have changed or are not available. If they are available, realize that some layouts and navigation tools change frequently. Be flexible, creative and, above all, patient when working with these websites. We have tried to use sites that are relatively stable and have provided the high-level URLs to each site's home page, along with directions on how to navigate to the individual data depositories (specific URLs change frequently). However, you may still find some sites are not available from time to time. If a particular site is not available, search for other sites that may host the same, or similar, data. You could try a different web browser, check your security settings, try again later, or contact the site administrator if the problem persists. If the interface has changed, use the site's search facilities to locate the resources you require. As an additional resource, a supplementary website to support this book will also be available, where recent updates and any major changes will be posted on a regular basis.

Summary and looking ahead

In chapter 1, we have introduced and defined public domain spatial data and discussed some of the licensing, access, and copyright issues associated with these data. By now you should appreciate that although a vast amount of information is publicly available, not all of it is in the public domain. This is an important distinction to keep in mind as you read this book and, more importantly, as you search for and use spatial data in your projects.

Chapter 2 begins with defining spatial data models and then focuses on vector data, including transportation, soils, hydrography, land parcel, agriculture, biomes, hazards, names, and demography. The section on data quality discusses some of the issues in defining what quality means.

Chapter 3 focuses on raster data portals, sources, and quality; in particular, elevation, land cover, imagery, climate, and population. It also examines some of the privacy issues associated with spatial data.

Chapter 4 begins with a focus on the true costs of spatial data and reviews the ongoing debate as to whether public domain spatial data should be free or fees charged for their use. The chapter will also consider whether spatial data should be subject to copyright and government policies, both existing and emerging.

How do data portals, national and state, and metadata standards affect your use of public domain spatial data? Chapter 5 focuses on these issues and explores the national spatial data infrastructure initiatives in a selected number of countries, providing a discussion on metadata and the issues surrounding data sharing.

Moving to the international scale is the focus of chapter 6, which looks at some of the issues involved and a consideration of whether we are moving toward universal access to global data.

By chapter 7, you will have covered most of the main issues and had enough practice with the exercises to "put public domain spatial data to work." In this chapter, you will examine some of your decisions as you locate and work with spatial data, and how to evaluate those decisions at each step of your project.

The last three chapters will focus on new initiatives that are already affecting GIS data and will have an increasing influence. Chapter 8 covers the dual role of data user and data provider. Chapter 9 discusses how the cloud computing revolution—something you have undoubtedly heard much about—is set to deliver the next generation of GIS data and services and what this will mean for data and GIS, with examples using ArcGIS online and elsewhere.

Finally, chapter 10 reflects on the future of public domain data, the emerging issues and initiatives, and how these will affect data quality, data availability, privacy, licensing, copyright, metadata, and user choice.

References

"Copyright Term and the Public Domain in the United States." 2011. *Cornell University*. http://www.copyright.cornell.edu/resources/publicdomain.cfm.

"Guidance for the Use of OneGeology Images and Data." 2011. *OneGeology*. http://www.onegeology.org/.

Harley, J.B. and David Woodward eds. 1987. *History of Cartography: Cartography in Prehistoric, Ancient, and Medieval Europe and the Mediterranean Vol. 1.* Chicago: University of Chicago Press.

"My IP Intellectual Property Explained." 2011. *Intellectual Property Office*. http://www.ipo.gov.uk/.

Patterson, Lyman R., and Stanley W. Lindberg. 1991.*The Nature of Copyright: A Law of User's Rights.* Athens, GA: The University of Georgia Press.

"Public Domain." 2011. *Wikipedia*. http://en.wikipedia.org/.

"Public Domain CC Wiki." 2011. *Creative Commons*. http://creativecommons.org.

"Public Domain Definition." 2011. *Linux Information Project (LINFO)*. http://www.linfo.org/.

"Public Domain Information." 2011. *United Nations Educational, Scientific, and Cultural Organization (UNESCO)*. http://portal.unesco.org/.

"Where is the Public Domain?" 2011. *United States Copyright Office*. http://www.copyright. gov/.

Wilson, J.D. 2001. "Attention Data Providers: A Billion-Dollar Application Awaits." *GeoWorld*. (February) p. 54.

2

SPATIAL DATA MODELS, VECTOR DATA, AND DATA QUALITY

Introduction

In chapter 1, you learned what public domain means, its effect on various spatial data sources that are increasingly available, and its impact on GIS practitioners. In the accompanying exercise, you downloaded public domain data to tackle global issues surrounding population, climate, and the environment. In this chapter, you will examine some of the main vector datasets available in the public domain and some of the issues associated with their use. The accompanying exercise is based on US Census demographic and street data to determine the best location for an Internet café.

As you saw in chapter 1, solving a problem using GIS and spatial analysis begins with the acquisition and investigation of spatial data. Most analyses involve some base spatial data—roads, hydrology, administrative boundaries, property outlines, and so on—that will be either central to your analysis or simply reference layers for data you generate yourself. As you work with GIS and spatial data to solve problems and make decisions, it is important to consider the intended use of the data, which influences the format you use. However, before learning more about the plethora of datasets available today, you should consider some issues that affect *all* data—no matter what format, no matter what source.

Spatial data models

Within a digital environment, graphically representing and interrogating real-world entities, such as school campus zones or the average path of hurricanes as they head west over the Atlantic Ocean, involves storing them within the framework of a spatial data model. The data model's coordinates define the geographic location and extent of each entity. It is this ability to manage location information, in either two- or three-dimensional space, that sets a GIS apart from other information systems. Another unique characteristic of a GIS is that the data model maintains not just the coordinates representing the location of each entity but also the relationships between entities. This is critical in examining issues such as:

> What is the relationship between birth rates and life expectancies between countries?
> How does the underlying geology affect the conductivity of streams?
> How do contour plowing techniques affect soil moisture and erosion on specific soil types and slopes?

Real-world entities like these are generally represented in a GIS using one of two data models—vector or raster.

Vector data model

Vector data models use discrete elements to represent objects as points, lines, or areas. This allows all positions, lengths, and dimensions to be defined precisely. As well as storing the x,y coordinate information describing the location of each entity, each vector type also stores related attribute information to identify, for example, what type of feature it is and how it should be displayed. In the vector model, collections of similar features (a feature layer) are stored as vector points, vector lines, or vector areas. A single

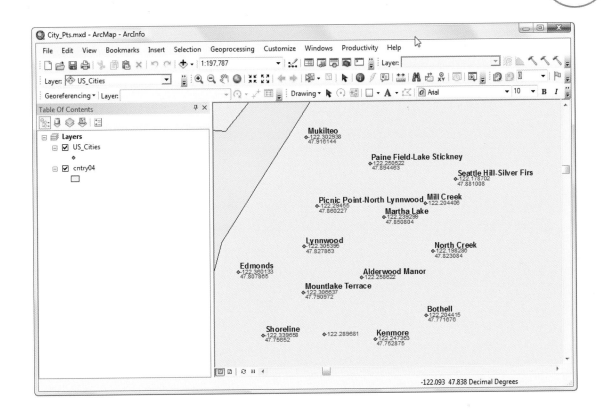

Vector point data: Cities, Washington State, United States. Each city has a unique latitude-longitude coordinate value, shown in blue.

feature layer never includes all three data types because individual entities are either points or lines or areas. Water wells, mountain peaks, and power poles are examples of entities typically represented as points. Railways, bird migration routes, and rivers are represented as lines, while census enumeration areas, wetlands, and geologic units are often represented as areas.

Map scale refers to the ratio of the distance on a map to the corresponding distance on the ground and is an important component in spatial data. For example, a scale of 1:10,000 indicates that one unit on the map (or in a GIS database) represents 10,000 units on the ground. One centimeter, or inch, or meter, on the map represents 10,000 centimeters, inches, or meters on the ground. As a paper or digital map at 1:1 scale would be too large and unwieldy, the earth is shown at a reduced scale in a map or GIS database and described by a ratio or representative fraction. All spatial data are collected at a specific scale. Spatial data at a large scale, such as 1:10,000—or written as a fraction, 1/10,000—show a great amount of detail but cover a small area. Conversely, spatial data at a small scale, such as 1:1,000,000, show a small amount of detail but cover a large area.

ORIGIN OF THE TERM

The term *large scale* was coined as the fractions representing large scale, such as 1/10,000, are larger numbers than those representing small scales, such as 1/1,000,000.

Scale has a significant influence on deciding which vector type best represents a particular entity. Cities may be better represented as points at a small scale but as areas at a larger scale, just as points best represent tornadoes at a small scale but at a larger scale, lines provide a better indication of the tornado's path.

Vector points

Vector points represent all geographical entities that are referenced by a single x,y coordinate pair.

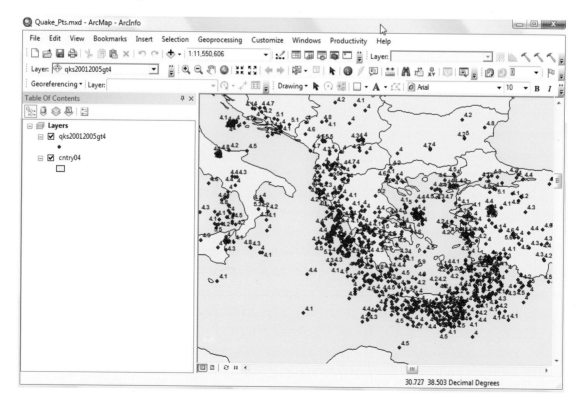

Vector point data: Feature annotation. Annotations indicating the magnitude for each earthquake have been created for each epicenter.

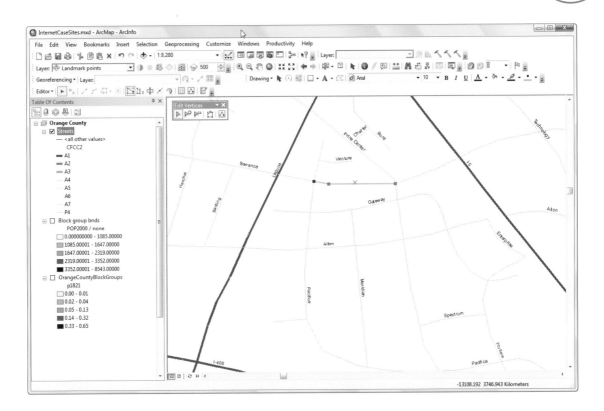

Vector line data: The vertices that comprise a single street segment in Orange County, California.

Vector lines

Vector lines representing linear geographical features are composed of a number of straight line segments, with each line segment defined by a number of vertices. The simplest linear feature contains a start and an end node (two x,y coordinate pairs). In ArcGIS, lines are referred to as *polylines* if they are stored as shapefiles and simply *lines* if they are stored in a geodatabase.

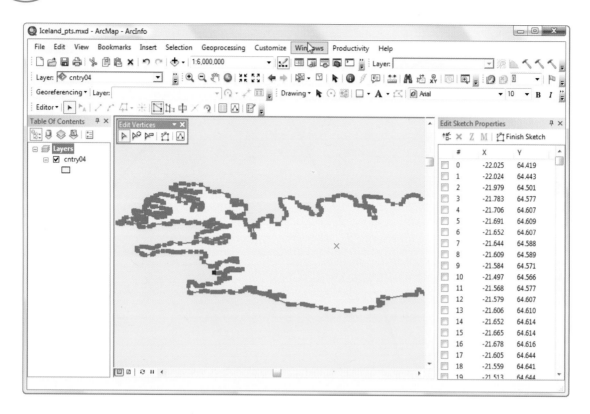

Areas such as the country boundary illustrated here, start and end at the same node.

Vector areas

Vector areas represent features that have a perimeter, or boundary, and an extent. The boundary is composed of line segments and x,y coordinate pairs, where the start and end node share the same geographic location. For topological consistency, to support near neighbor, point-in-polygon searches and so on, the line segments must not overlap or intersect.

Raster data model

In the raster data model, objects are represented by cells or pixels. A simple raster data structure consists of a 2D array of grid cells. Each grid cell is referenced by a row and a column number and contains a number representing the type (value) of the attribute being mapped. For example, in a raster dataset representing snow cover, a pixel with a value of 0 may represent no snow and a value of 1 represents areas of snow. As each cell can only contain one number or attribute, different attributes must be represented in separate 2D arrays (or overlays). Most raster data analysis involves combining 2D arrays to create new layers and cell values (Burrough and McDonnell 1998).

Raster data models are either *discrete* (thematic) or *continuous*, depending on what they represent. A forestry land use raster with a value of 1 that indicates deciduous trees, a 2 representing coniferous trees, and a 0 representing no trees is an example of a single-layer thematic raster, as the cells are encoded with a value from a discrete range of possible values. An elevation grid, or digital elevation model (DEM), is an example of a single-layer continuous raster where each cell is encoded as a floating point decimal number and each cell has an infinite number of potential values. Satellite imagery, such as the multi-band imagery from the Landsat Thematic Mapper satellite, is an example of a multi-layer continuous raster format; cells could have an infinite combination of reflectance values in different bands of the electromagnetic spectrum.

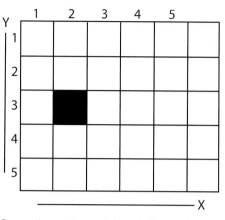

Raster points are represented as a single grid cell.

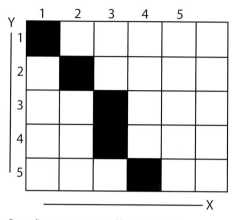

Raster lines are represented by a number of neighboring cells stretching out in a given direction.

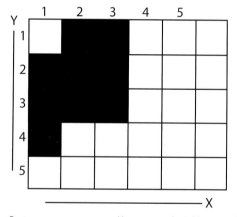

Raster areas are represented by a group of neighboring cells.

Although hybrid data models such as the object data model and triangulated irregular networks (TINs) do exist, most public domain spatial data are available in either vector or raster format. Certain features such as water wells, trails, and municipal land use zones are typically represented as points, lines, and areas (respectively) in the vector model, and other characteristics of the earth such as elevation or land cover are represented in the raster data model. However, it is possible to convert data from one data model format to another. For example, point vector data representing soil pH could be converted to a raster grid for the purpose of estimating, or interpolating, what the pH would be in areas not specifically sampled. The same data could be visualized in 3D, with mountains representing areas of high pH, valleys representing low pH soils, and cliffs representing significant changes over short distances, which merit further investigation.

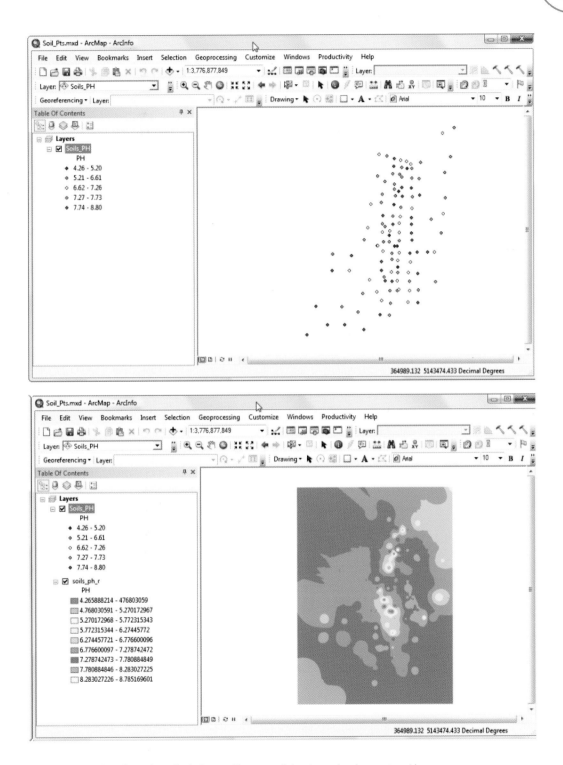

Top: Point vector data of pH values of soils. Bottom: The same soil data interpolated as a raster grid.

However, even though data can be easily converted from one data model to another, converting between vector and raster formats could influence the final results of the analysis, as we will see in the exercise for chapter 3, and may compromise the accuracy of the data.

Data format and availability

Spatial data typically comprises multiple files rather than a single file because GIS is a *system*, and each file plays a certain role in that system. Some files deliver the G (geography) parts of GIS (the map or image), while others provide the I (information) parts of GIS (often in associated attribute tables), and the remainder deliver the S (system) parts of GIS (the spatial information or the topological relationships between features). As real-world features are complex, the files containing the data that represent them tend to be quite large. As a general rule, for a given geographic area, raster format datasets tend to be larger than their vector counterparts, a point worth remembering when bandwidth restrictions are an issue. Consequently, spatial data are typically compressed and made available as a small number of archive or zip files. This has the twofold benefit of minimizing the total number of files that the end user must download while reducing the volume of data to be transferred across the Internet.

Besides their accessibility as downloadable files, public domain spatial datasets are also increasingly available as data services. A data service provides access to spatial information that resides on a remote server, along with some tools that allow users to interact with and interrogate that information. These data services may be accessed by either downloading or streaming the data to a desktop application such as ArcMap or ArcGIS Explorer. (Streaming the data involves the transmission of digital data from one computer to another without having to store it at the destination). Spatial data services are developing rapidly, both in terms of the number of services that are available and the increasing sophistication of the accompanying analytical tools. This is having a profound impact on both the organizations that provide the data services and the spatial data user's experiences and expectations, as we shall see later on.

Vector public domain data

So, you have a spatial problem to solve, you have determined the requirements of your project, and you have identified a need for some public domain vector data. As you know from chapter 1, you can obtain public domain spatial data from a variety of international, national, state, and local portals hosted by government agencies, non-profit organizations, universities, and private enterprises. You also know that a large volume of data exists, much in vector format.

However, prior to working with that vector data, and to avoid any unnecessary data collection, you should first be aware of what types of data are already available so you can identify and download the most appropriate dataset(s) for your particular project. There's little point in embarking on a costly data capture exercise if it's not absolutely necessary. Second, most phenomena that GIS analysts seek to model and analyze do not exist in a vacuum; they are interconnected with other phenomena, often in complex spatial relationships. For example, the traffic in one segment of highway affects and is affected not only by the adjacent highway segments, but also by segments that may be dozens of kilometers distant. Similarly, flooded rivers are influenced by the shape, extent, and complexity of the entire watershed; it's not just the immediate section of river under flood that requires investigation. As such, bear in mind that you may need to consider downloading a larger amount of data, or additional datasets, to successfully complete your project.

The next section reviews some main categories and sources of global and regional spatial data considered core to GIS analysis.

Key public domain global vector datasets

The Digital Chart of the World (DCW) was arguably the first global vector public domain dataset. It was produced during the late 1980s from the 1:1,000,000-scale US Defense Mapping Agency's Operational Navigation Charts. The data included vegetation, transportation, cultural landmarks, land cover, drainage, utilities, airports, railroads, roads, land and ocean features, populated places, and country boundaries. Last updated in 1992, it was originally divided into 2,094 tiles representing 5 degree × 5 degree tiles for the world and was subsequently modified to make it more easily readable by modern GIS software. It had been made available via the Pennsylvania State University Libraries website, but this service has now been withdrawn due to inaccurate and out-of-date information. Prospective users are re-directed to the Natural Earth Data website.

The Vector Map (VMAP) dataset was created from the DCW and is available at three resolutions.

Low resolution: VMAP0 (1:1,000,000 scale) represents an update of DCW, is distributed in four regions of the world, and is in the public domain.

Medium resolution: VMAP1 (1:250,000 scale) is divided into 234 tiles, only a fourth of which are available for download from the US National Geospatial-Intelligence Agency (NGA) distributed by MapAbility. The boundaries data and the reference library are copyrighted to Esri.

High resolution: VMAP2 (1:50,000 scale) is not currently available for public release.

Much of VMAP's data are available in NGA's Vector Product Format (VPF). This format can be read by ArcGIS Explorer, but to use it with ArcGIS requires that the data be converted using a utility such as Global Mapper. Alternatively, companies such as Global Mapping International have converted the VMAP0 data into ready-to-use GIS layers for a fee.

Another set of world base data is Global GIS Data. This dataset genuinely lives up to its name, with twenty-eight vector layers and eight raster layers, including elevation, land cover, earthquakes, river basins, geologic province, ecological regions, vegetation "greenness" images, and more. The US Geological Survey (USGS) and the American Geological Institute (AGI) collaborated to make it available as a world atlas, comprising USGS and other public domain data at a scale of 1:1,000,000 in Esri GIS-compatible format. The dataset is produced by region, each being defined as rectangular blocks bounded by lines of latitude and longitude. It is not available for download but may be ordered as CDs or a DVD on the AGI website. Accompanying metadata are available from the USGS Global GIS Project website or from the AGI site.

Global Map is an international effort involving a number of national mapping organizations to map the world, at a common scale and resolution, using the same data dictionary. The scale for vector data is 1:1,000,000 and the resolution for raster data is one square kilometer. Japan leads the Global Map effort and Esri, through a grants program, provides GIS technology to some of the participants. The four vector layers are boundaries, transportation, drainage, and population centers, while the four raster layers are elevation, vegetation, land use, and land cover. Only two raster layers, discussed further in the next chapter, are complete at the present time. The Global

Map Data may be downloaded from the International Steering Committee for Global Mapping website.

The largest single source of spatial data, some in the public domain, is probably Esri's Data and Maps for ArcGIS. The Data and Maps for ArcGIS five-disc set contains over 26 gigabytes of data from a myriad of public and private sources. The Data and Maps for ArcGIS StreetMap North America DVD contains the StreetMap North America dataset from Esri and Tele Atlas, as well as Data and Maps for ArcGIS vector data for the world. The other four DVDs contain elevation and image raster datasets. Disc 2 contains the 90-meter Shuttle Radar Topography Mission (SRTM) global DEM along with other worldwide elevation and image datasets, and discs 3–5 contain regional data for North and South America, Europe and Africa, and Asia and Australia. Each of these regional DVDs includes global imagery captured at 150-meter resolution, as well as shaded relief derived from the SRTM global DEM. All five DVDs contain the Data and Maps for ArcGIS HTML-based Help system.

Making the most of data sources like the Data and Maps for ArcGIS DVDs and the Global GIS CDs can significantly reduce the amount of time users have to spend searching for, downloading, and manipulating data directly from both public domain and private data sources online. The vector data on the Data and Maps for ArcGIS DVDs are stored in the Smart Data Compression (SDC) format created by Esri to reduce storage requirements and be directly readable in ArcGIS for Desktop software. Esri collated the data from a wide variety of sources, packaged them in formats ready to be used within ArcGIS, and provided the necessary supporting documentation, or metadata. The metadata conform to both the Content Standard for Digital Geospatial Metadata (CSDGM) from the Federal Geographic Data Committee (FGDC) and the International Organization for Standardization (ISO) 19115 standard. To make the metadata more accessible and useful, Esri defined additional elements to support automatic metadata updates and to document additional characteristics. These elements are defined in the Esri profile of the CSDGM.

The data on the Data and Maps for ArcGIS DVDs may be used in an infinite variety of projects, but like other datasets that have been augmented by private vendors, much of the data cannot be redistributed because not all of the data originally came from public domain sources. A matrix indicating which datasets may be redistributed, along with the associated terms and conditions, is available via the ArcGIS web-based Help system. Some data have no redistribution rights and so may only be used for specific internal use. For other data, you have redistribution rights as long as you provide full metadata and source or copyright attribution to the respective data vendors. Some data are only redistributable with a value-added software application developed by Esri partners, again with full metadata and attribution, while other data are redistributable without a value-added software application with full metadata and attribution. The last category is public domain data from the US government, freely redistributable with appropriate metadata and source attribution.

Hydrographic data

Hydrographic (or hydrologic) data contain information on lakes, shorelines, rivers, and sometimes on human-made water features such as canals, ditches, and reservoirs. As with other data themes, small-scale hydrographic data sources are available for the planet. The first global hydrographic dataset, HYDRO1k, was created during the late 1990s. It has a horizontal resolution of one kilometer and is based on elevations in the GTOPO30 dataset (the Global 30 Arc-Second Elevation Dataset available from the USGS, which will be discussed further in chapter 3). GTOPO30 contains both raster and vector data, including rivers, catchment basins, slope, and aspect (direction of slope).

The Global Lakes and Wetlands Database (GLWD) from the World Wildlife Fund website is another comprehensive global hydrography dataset that includes extensive attribute data. The scale of the data ranges from 1:1,000,000 to 1:3,000,000 and is available on three levels:

- Level 1 (GLWD-1) comprises the shoreline areas of the 3,067 largest lakes (area at least 50 km^2) and 654 largest reservoirs (with a storage capacity of at least 0.5 km^3).
- Level 2 (GLWD-2) comprises the shoreline areas of permanent open water bodies with a surface area at least 0.1 km^2, and includes 250,000 areas as lakes, reservoirs, and rivers.
- Level 3 comprises lakes, reservoirs, rivers, and different wetland types in the form of a global raster map at a resolution of 30 seconds. GLWD-2 and GLWD-3 do not provide detailed descriptive attributes such as names or volumes.

Global and medium-resolution hydrography datasets—national and regional maps showing rivers, lakes, and even watersheds—have existed for decades in many parts of the world. As yet, no seamless high-quality large-scale hydrographic data exists for the planet. Reasons for this include the difficulty of surveying and mapping rivers due to their dynamic nature and the political sensitivities associated with rivers that form administrative boundaries. Other factors hindering the comprehensive detailed mapping of large river basins are their often remote locations and the fact that they may transect a number of countries (Lehner et al. 2008). Recent flood events in Brazil,

Hydrographic data describe the spatial extent, and sometimes the condition, flow direction, and other attributes of flowing and stagnant water bodies. Photo by Joseph Kerski.

Australia, Sri Lanka, and Thailand have once more highlighted the urgent requirement for large-scale hydrographic datasets to support the holistic and systematic study of watersheds, water quality, environmental change, and flood hazards.

In the United States, the original hydrographic data were derived from USGS topographic maps, digitized originally as digital line graphs (DLGs) during the 1980s and 1990s. During the same time, the US Environmental Protection Agency (EPA) developed a database, known as the Reach File, which recorded unique identifiers for 3.2 million stream segments or *reaches*. A reach is a segment of a river or coastline considered to be homogeneous according to the EPA classification method. The three versions of the Reach File that currently exist, RF1 (1:500,000), RF2, and RF3-Alpha respectively (1:100,000), were created from detailed sets of hydrography data produced by the USGS and are available for download from the EPA website.

Each of these datasets provides different advantages. The DLGs were spatially accurate while RF3 contained the hydrologic network, names, and links to the thousands of stream gauging stations maintained by the USGS. During the last decade, the National

Hydrography Dataset (NHD) was developed as a combination of the best of the DLGs and RF3. NHD, available to download from the USGS website, contains not just streams, but also lakes, reservoirs, canals, pipelines, dams, and other features. These features were initially provided at a 1:100,000 scale, but are now available at the larger 1:24,000 scale. Unlike most vector datasets, NHD is a networked dataset in which the network is based on the geometry and connectivity of the individual features. As such, it may be used in conjunction with Esri's ArcGIS Network Analyst extension or the Utility Network Analyst tools to determine what is connected—upstream or downstream —from a desired point along a stream, or what rivers form a certain watershed. This connectivity is essential for hydrologic analyses. For example, if a truck carrying hazardous liquids overturns and spills toxic chemicals into a nearby river system, analysis of the river network requires the rapid identification of all the reaches downstream that would be affected.

An enhanced 1:100,000-scale database, NHDPlus, contains value-added attributes to enhance stream network navigation, analysis, and display. It also contains an elevation-based catchment for each line in the stream network, catchment characteristics, headwater node areas, cumulative drainage area characteristics, flow direction, accumulation, and elevation grids. Each flow line in the stream network includes flow line minimum and maximum elevations, slopes, and flow volume and velocity estimates.

The World Wildlife Fund, USGS, and McGill University collaborated on an improvement to HYDR01k, called "Hydrological data and maps based on Shuttle Elevation Derivatives at multiple Scales," or HydroSHEDS. Based on 3-arc-second SRTM data, a DEM for the planet, HydroSHEDS has a more detailed horizontal resolution than HYDR01k. Unfortunately, only the DEM and the flow direction grids are available at the full resolution at this time. GIS analysts are able to use the flow direction grid, along with their own GIS and modeling software, to develop flow accumulation layers and flow lines. Stream and drainage basin shapefiles are available at 15- and 30-arc-second resolutions, but unfortunately no funding is currently available to complete the derived products (similar to those created for the HYDR01k dataset) at the full resolution scale. HydroSHEDS data are available from the USGS.

Outside the United States, some larger scale hydrography data are available in the public domain courtesy of the national mapping or environmental agency. Surface water has long been a standard feature on national topographic maps, and it is from these features that most hydrographic data have been derived. Hydrography data are generally available as a set of one or more vector layers that are not vertically or horizontally integrated with other layers. Occasionally the data may be available in a networked vector format; as rivers are networks by their very nature, the network vector format is the preferred format. In 1992, UK Hydrographic Office and Ordnance Survey joined forces to produce a prototype coastal zone map for the United Kingdom. Since then, the UK Hydrographic Office has converted its holdings to the Admiralty Vector Chart Service. This data format is designed to be used as a stand-alone digital map product rather than tied to any specific GIS.

The Australian Hydrographic Service operates a node of the Australian Spatial Data Directory (each node represents an online collection of metadata documents) and offers several types of spatial hydrographic data online. The most suitable for GIS is the Seafarer GeoTIFF product, georegistered raster images of Australian Navigational Series maps. Anyone wishing to use the data must generate their own hydrological network from the individual files.

Elevation data

Throughout the twentieth century, many national mapping agencies around the world developed topographic maps that were primarily for the scientific community. The most time consuming part in creating these topographic maps was the contour lines. They were meticulously etched onto copper plates until mid-century, then on plastic scribecoat film from the 1950s through the 1980s, and subsequently digitized using photogrammetric and GIS-based technologies. Although a number of private companies have been involved in creating topographic maps as well, their mapping efforts have typically been focused on much smaller geographical areas, such as construction sites, where a higher level of detail is required.

The contour lines created by federal agencies, with an interval anywhere from five to a few hundred feet or meters, became the source data for most early DEMs. Most global elevation data, such as the data on the Global GIS CDs, are in DEM raster format. This is not a significant problem for anyone wishing to use the data because contour lines can be easily generated in ArcGIS from DEMs. However, contour lines generated in this manner are interpolated positions, at times from DEM cells that are 30 meters or even one kilometer on each side, and as such will be less accurate than contour lines drawn from the interpretation of stereo image pairs via photogrammetric methods.

In the United States, a few 1:24,000-scale DLGs contain contour vector data at 5-, 10-, 20-, or 40-foot intervals, depending on the terrain, while the 1:100,000-scale DLGs include national coverage at intervals of 100 meters or more. DLGs and other USGS data are available from the two of the three most commonly used portals from

Relief on the landscape can be modeled as vector data (contour) or raster data (grids) in spatial datasets. Photo by Joseph Kerski.

that agency, the USGS Earth Explorer and the USGS Earth Resources Observation and Science (EROS) Data Center. The third portal, Seamless Data, will be discussed in chapter 3.

Contour lines as vector data are provided by other national mapping agencies. These include contour lines at an interval of 1,000 meters from the Mexican census and mapping agency, the Instituto Nacional de Estadística y Geografía (INEGI). In Britain, contour lines are available from Ordnance Survey in the form of their OS Landform Profile or OS Landform Panorama data products. Landform Profile products are derived from 1:10,000-scale maps and are updated frequently. They are delivered as standard and as index contours, with a vertical interval of 5 meters, except in mountain and moorland areas where the interval is 10 meters. The index contours are every fifth contour. The Landform Panorama contours are generated from 1:50,000-scale maps with contours at 10-meter intervals. Land Information New Zealand provides access to downloads of small areas via their LINZ Topographic Maps service.

Wetlands have only recently been valued enough to collect as spatial datasets. Photo by Joseph Kerski.

Wetlands data

For much of human history, wetlands were considered nuisances; hindrances to travel and a breeding ground for mosquitoes, snakes, and other unwelcome creatures. They were deemed useful only if they could be drained and developed for settlement or agriculture. Consequently, wetlands data may not have been collected on the first topographic maps produced in the 1800s. It is only in the past century that wetlands have come to be appreciated for their roles in flood control, biodiversity, habitat, recreation, and much more.

Some examples of wetland datasets include the world wetlands data from the WWF (World Wildlife Fund) GLWD mentioned previously and the global dataset, at 1 × 1 degree resolution, produced by the NASA Goddard Institute for Space Studies. These data cover five categories of wetlands: forested bog, nonforested bog, forested swamp, nonforested swamp, and alluvial formations.

Some wetland data are available at larger scales for some countries and local areas. The National Wetlands Inventory (NWI) is the key national source of wetland information in the United States. The data were originally obtained from USGS topographic quadrangles and have been updated by subsequent field surveys and through some analysis of aerial and satellite imagery. Maintained by the US Fish and Wildlife Service, NWI data are available by state through an Open Geospatial Consortium (OGC) compliant web mapping service. With each zipped file downloaded, the user receives wetlands data as vector-based area shapefiles, the accompanying metadata, and historic map report information. Using the wetlands mapper tool, users can extract wetlands data within the current view extent, but only one data layer at a time. Wetlands data can also be viewed in Keyhole Markup Language (KML). NWI data are being used to develop a Wetlands Master Geodatabase that will ultimately provide improved data to end-users, as well as provide land managers with the ability to manage these fragile environments more effectively.

At the state and local level, government agencies sometimes compile wetlands data into a downloadable database. Examples include the Arkansas Wetland Resource Information Management System, the North Carolina Department of Environment and Natural Resources website for the state's counties, and in Colorado, the Boulder County website.

Agriculture data

Agricultural statistics and data have been collected for centuries and remain one of the most published, and publicly available, datasets in the world. Agricultural data includes detailed information such as crop types, crop acreage, value, and total harvest at scales ranging from individual farm units to county-level studies to entire countries. Although online mapping tools, such as Crop Explorer from the Foreign Agriculture Service of the US Department of Agriculture (USDA), can analyze crop patterns, these mapping tools typically cannot support more detailed data analysis. The service does, however, provide some data in the form of Comma-Separated Values (CSV) files that can be downloaded into a desktop GIS for further analysis. (Search for "Downloadable datasets.") The most efficient way to study agriculture data is to download and analyze the data within a full-featured GIS.

The Food and Agriculture Organization (FAO) of the United Nations provides one of the most comprehensive global datasets on agriculture. Although much detailed data are available from the FAO Statistical Database (FAOSTAT), this site illustrates some of the problems working with agricultural data, in particular the requirement to manipulate the data in some manner before it can be used within a GIS environment. Detailed information by country and by crop is available, but as customized tables of data. These tables must be stored separately and then joined to a shapefile or geodatabase for mapping and analysis within ArcGIS. As another example in the United States, the National Agricultural Statistics Service is a rich data resource provided by the USDA. However, the data are available in text and PDF formats, which require some manipulation to get them into a format suitable for GIS.

One of the few agricultural datasets easily imported into and analyzed in a GIS is provided by the National Atlas. This provides data on crop type, value, amount of area in crop, and other agricultural data by state and county for the United States. The National Atlas team has modified the agricultural census data before posting the data to the website, thus adding value for the end-user and reducing the time required to reformat the data.

Boundaries and land parcel data

National, regional, and local records of land ownership and parcel boundaries, referred to as cadastral data, are among the oldest forms of geospatial data. Some of these inventories are centuries old, such as the Register of Sasines in Scotland, which dates back to 1617. Sasines are legal documents that record the transfer of property or land ownership. In 2007, the oldest surviving deed for land on Long Island, New York was auctioned for $156,000. The document was signed by Dutch Colonial Governor Wouter von Twiller at Eylandt Manhatans on June 6, 1636, and recorded the purchase of 3,600 acres from the Lenape Indians.

Land ownership and parcel boundaries are also among the most elaborate types of spatial data. In many cases, the boundaries between countries represent a complex and intricate history of territory won and lost. These often convoluted entities change over time and may be disputed by countries on either side. The situation may become even more complex at the subnational level, with administrative areas known around the world by a variety of names, such as provinces, states, departments, *bibhag*, *bundeslander, daerah istimewa, fivondronana, krong, landsvæðun, opština, sous-préfectures,* counties, and *thana*.

Land parcel boundaries, as here in Bavaria, Germany, are one of the oldest categories of features mapped, and yet they are still difficult to obtain as GIS-ready datasets. Photo by Joseph Kerski.

Cadastral data are available in a wide variety of formats and styles, reflecting how different cultures value and record such data. Many countries lack parcel boundaries or street addresses; some cultures do not consider land to be something that can be owned and sectioned off. Consequently, a truly global land parcel dataset cannot be generated at this time. If it could, it would almost certainly be one of the largest, and possibly one of the most complex, datasets ever created. Instead, the GIS practitioner is confined to using land and cadastral data at the country, regional, and local level. However, even at these scales, very little of the data is in the public domain for legal and privacy reasons.

Probably the most comprehensive database of administrative areas is the GADM database of Global Administrative Areas, a collection of 226,439 administrative areas in shapefile, geodatabase, or KMZ formats. Another source is the Second Administrative Level Boundaries (SALB) dataset project launched in 2001 by the United Nations Information Working Group. The data are maintained at a scale of 1:1,000,000 and subject to some United Nations copyright restrictions.

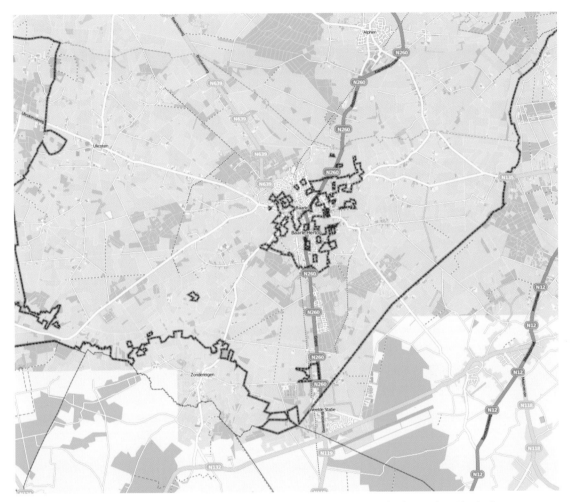

An example of a notoriously complicated boundary is the border between the Netherlands and Belgium, where islands of Belgium appear within the extent of the Netherlands and vice versa, and even islands of the Netherlands inside the Belgium islands that are inside the Netherlands! In this map, the community of Baarle-Nassau is inside the Netherlands but is partly comprised of territory that is part of Belgium. Esri World Street Map layer in ArcGIS Online.

The boundary between the Netherlands (left) and Belgium is marked with crosses on the ground along the side of a café. Photo by Jérôme Kunegis.

Some private companies maintain high-resolution world basemaps containing boundaries, such as the DeLorme World Data Base, a global map with a horizontal accuracy of 50 meters. This dataset, and others like it are available for sale and subject to copyright restrictions. However, Esri is processing a lower-resolution version of the DeLorme map for release as a layer package downloadable on ArcGIS Online.

INVESTIGATING LOCAL CADASTRAL DATA

Visit the online City of Greeley, Colorado, Online Resource for Interactive Greeley INformation (ORIGIN) Property Information Map system. This mapping system, based on ArcGIS for Server, has been available since 2009.

What information is contained in Greeley's online mapping database?

How much information on individual land parcels is available?

What layers do the city use for base mapping?

Where did these layers come from?

How many of these layers do you think are public domain?

In the United States, the GeoCommunicator service from the Bureau of Land Management (BLM) contains the land survey grid for all BLM lands and for some additional territory. This includes the public land survey system, which covers most of the country from Ohio westward, begun with the Ordinance Act of 1785. With this act, the government addressed the settling of the interior by dividing up the land into townships, ranges, and sections, and subdivisions of the sections. Individual landholdings were based on this system, but only in the states where the system was in place, and parcels do not always adhere to the system. Township, range, and section lines are included in GeoCommunicator, as well as some survey markers such as benchmarks, section corners, and triangulation stations. Cadastral data are increasingly available for browsing via web mapping services hosted by city and county governments, but are seldom downloadable as a public domain data source from those agencies due to restrictions on its use.

Although obtaining land parcel data from multiple local government organizations can be a daunting task, there are recent initiatives to improve access to this critical data resource. A spatial data provider, eMap International, announced during 2009 that the First American Corporation was developing the first ever national parcel database. This vector area dataset contains the boundaries of each parcel, with attributes for assessor parcel number, address, and owner name (if available). Data are sold by the county, with varying prices (from $900 to over $30,000 for a perpetual license) for the data to reflect the differences in detail and size of each county. By mid-2009, eMap International was offering access to land parcel data for 80 percent of the counties in the United States. Although most of the data have been obtained from individual county governments, some data were created from paper maps for those counties where digital data were unavailable. The land parcel database is primarily marketed to urban planning, architecture, oil, gas, renewable energy, engineering organizations, and industries. Interestingly, eMap International initially made its name by reselling Digital Globe and other satellite imagery. Given the fact that its president and CEO is not only a certified photogrammetrist, but also a professional land surveyor, it is not surprising that eMap invested a great deal of time and effort in compiling parcel and other vector data, as well.

Buildings data

As buildings are such small features on a global scale, and ubiquitous in many areas, they typically only appear on certain large scale products produced by national mapping agencies. Building footprints—areas that represent building location and extent— have long been included on Ordnance Survey maps in Britain and on USGS maps in the United States. Footprints are a part of the OS MasterMap dataset, maintained in a topography layer which includes not only contours and other features of the natural landscape, but nearly a half a billion features from the built landscape as well.

In the United States, buildings data were considered a lower priority for digitization by the USGS than other data such as hydrography and roadways, and consequently, buildings are only available as a DLG layer in very limited areas. Regional and local governments almost always maintain buildings data in their GIS databases due to the priority for producing tax assessor maps. Buildings, along with attributes such as the owner, building type, area, and date constructed, are occasionally available to download from these organizations. You will explore one of these buildings layers for the City of Boulder, Colorado, in the exercise that accompanies chapter 6. Partly as a result of the lack of buildings and other local infrastructure data, communities and universities are being invited to contribute their data to the Esri Community Maps program,

University buildings, such as this one on the campus of the Massachusetts Institute of Technology, are some of the most admired structures in the world. However, spatial data about individual buildings remain elusive. Photo by Joseph Kerski.

to build a global high-resolution dataset that includes buildings and other detailed features. This initiative will be discussed further in chapter 9.

Environmental data

Environmental spatial data are increasingly available as countries and communities realize just how vital this information has become in addressing global environmental issues. Some global-scale environmental datasets do exist, and there are a number of initiatives for collating the data gathered by individual countries into larger datasets. One of the most comprehensive spatial data libraries of environmental information is the International Union for Conservation of Nature's Red List of Threatened Species. Information on approximately 28,000 of the 56,000 assessed species include spatial data (as shapefiles) containing areas depicting the ranges of the species and associated tabular data containing taxonomic, distribution, and source information.

The European INSPIRE directive (Infrastructure for Spatial Information in the European Community) stems from the growing awareness of the need for access to better information, to understand the interactions between human activity and environmental impact. The directive aims to promote better decision making and a coordinated

Environmental spatial data represent some of the most valued spatial data of all, but there is often a lack of consensus as to what should be included in these datasets. Photo by Joseph Kerski.

response to environmental issues through the sharing and exchange of data between government agencies in European Union countries. INSPIRE will establish an infrastructure to deliver integrated, georeferenced spatial information services to all member states. We will return to INSPIRE in chapter 6 when we investigate international data portals.

The WWF's comprehensive database of 827 terrestrial ecoregions from Olson and colleagues (2001) has found its way into numerous web mapping tools, and is available to download. More recently, WWF released the Global 200 dataset, containing ecoregions whose conservation "would achieve the goal of saving a broad diversity of the Earth's ecosystems" (search for WWF's Global 200). Floramap, from the Centro Internacional de Agricultura Tropical in Colombia, is a set of GIS-compatible maps that show the most likely distribution of wild species in nature. The center also hosts climate data, a combination of temperature and rainfall with weightings adjusted for latitude, in the form of grids for different regions of the world.

In the United States, the EPA's GeoData Gateway provides access to thousands of points in the toxic release inventory, other pollutants, conservation areas, and basemaps. On a continental scale, the North American Environmental Atlas contains data on watersheds, terrestrial ecoregions, human impact on protected areas, industrial pollution, and a base set of transportation, waterways, and cities layers. The atlas was produced as a result of collaboration among the national mapping agencies of Canada, the United States, and Mexico through the Commission for Environmental Cooperation. The site is one of the more GIS-user friendly resources found online, as the data are offered as shapefiles and layer packages; even a map document is available. Metadata and KML files are also readily available and the site provides a viewer for examining the data before downloading. Global Map, a resource mentioned earlier that may be considered environmental in scope, represents another collaborative effort among national mapping organizations to combine their individual output into something that would benefit global research.

Energy data

Data on the location of oil, coal, natural gas deposits, power plants, and other energy-related features are, as you might expect, extremely difficult to obtain. This is primarily due to security concerns and the commercial sensitivity surrounding the location and extraction of energy resources. The combination of fierce competition between private energy companies and associated security concerns means little, if any, energy data are available in the public domain. While some citizen watchdog groups provide regional locations and lists—for example, Appalachian Voices commissioned and maintains a resource of mountaintop removal data for the Appalachian Mountains—the coverage of such lists is often limited to the organization's region of concern. At the international level, the USGS Energy GIS Data Finder provides a portal to all free GIS datasets from the various energy programs. These are often the result of studies covering localities or regions, such as coal lease areas in the Danforth Hills of Colorado or the North Sakhalin system of Eastern Russia. However, there are several key global datasets, such as the USGS World Petroleum Assessment, which includes GIS datasets and interactive maps. Another key resource is the International Energy Agency, which hosts statistical data on energy consumption, taxes, and output. Some of their data are available for sale and some are free. Most of their data are available in PDF or Microsoft Excel format, requiring some additional processing for GIS use.

In the United States, the National Renewable Energy Laboratory (NREL) offers data by state or for the entire country on such topics as wind, solar, and geothermal energy. The Energy Information Administration from the US Department of Energy includes

GIS databases on consumption, prices, expenditures, refineries, power plants, coal mines, distribution, and renewables available. The USGS National Oil and Gas Assessment provides access to domestic petroleum assessments performed by the agency's energy program.

Many state oil and gas associations in the United States also host data online, although the data are usually not as current as those held by private companies. The available data ranges from those already in GIS format, for example from the Kansas Geological Survey website, to data that must be queried and manipulated to format for GIS, such as the data available from the Utah Department of Natural Resources Division of Oil, Gas, and Mining. Among the most comprehensive resources are the energy data from the Colorado Oil and Gas Conservation Commission.

The Rocky Mountain Oilfield Testing Center (RMOTC) maintains a detailed set of 2D and 3D GIS oil and gas layers for its test site in Wyoming. RMOTC's spatial data are not typically made available for general use but are available on request for educational and research purposes.

Geologic data

Ever since the world's first geologic map was made by William Smith, who tramped about England and Wales during the 1810s observing and gathering data on the variations he encountered in rock strata and structure (described in the book *The Map That Changed the World*), people have attempted to map what couldn't be seen underground. As many events and phenomena on the earth's surface are affected by what lies below, GIS analysts often need information on surficial and bedrock geology, fault lines, mineral deposits, and mines. Keep in mind that, although much of these data came from paper maps drawn according to meticulous mapping standards, nobody was underground measuring the exact location of the rock formations and geological phenomena. All geologic data are estimates based on drill hole logs, seismic experiments, rock samples, and other measurements. Although the techniques for identifying geological features have become increasingly sophisticated since Smith's time, the location and extent of many rock features are still based on estimates between and outside the locations of primary sampling. Producing a geologic map today no longer takes the twenty years it took Smith to complete his map, but it remains expensive and time consuming.

That said, some small-scale global geologic data do exist. OneGeology, mentioned briefly in chapter 1, is a flagship project of the International Year of Planet Earth. It has been billed as the world's largest geological mapping project. Supported by

Geologic data ranges from simple faults, geologic units, and outcrops to detailed information about strike-and-dip, subsurface, and surficial layers, and much more. This is a basaltic outcrop in Connecticut. Photo by Joseph Kerski.

UNESCO and nine other international umbrella bodies, its aim is to create a dynamic 1:1,000,000 geological map data of the world available via the web. At the initial meeting in the United Kingdom in 2007, eighty-one participants from forty-three nations and fifty-three national and international bodies asserted that geological map data are essential to advancing science and education. Having access to such data would provide solutions to the challenges of mitigating environmental hazards, ensuring a sustainable supply of energy, minerals, and water, and addressing the urgent challenge of climate change. As of early 2011, 117 countries were taking part in the project. Data from OneGeology may either be visualized via a web mapping service or downloaded as a KML and viewed in such packages as ArcGIS Explorer or Google Earth, or converted into Esri shapefile or geodatabase format.

Another source for regional and global geology, as well as geochemistry, geophysics, and mineral resource maps and data, is the USGS Mineral Resources On-Line Spatial Data library. Some data are available for download, while other data are available through OGC compliant map services. Although this resource is hosted by the USGS, other federal, state, and local government agencies are invited to contribute data to its library. Examples of data available in the library include magnetic anomaly maps, gravity anomaly maps, geochemistry of stream sediments, geologic maps, and mines and mineral plants. Some of the data are in geodatabase or shapefile format, while others are in spreadsheet format, delimited text format, or raster grid. A key component of the library is the Mineral Resources Data System which contains hundreds of thousands of records, providing information on metallic and nonmetallic mineral resources throughout the world, including deposit name, location, commodity, deposit description, geologic characteristics, production, reserves, resources, and references.

At the national level in Britain, the British Geological Survey (BGS) provides 1:625,000-scale geologic data online in ArcGIS, MapInfo, and KML. In late 2009, the BGS opened a new OpenGeoscience portal, which provides access to its major data holdings using a map interface, along with a range of educational and professional tools, and the BGS's extensive national archive of geological photographs.

In the United States, the USGS created the country's first geologic maps for states and regions. They were also occasionally generated at the 1:24,000 quadrangle scale, the same scale as the largest scale of topographic maps. However, this type of mapping has traditionally been funded for local, one-time projects, rather than as part of a national geologic mapping program. As a result, the 1:24,000-scale geologic quadrangle data are patchy, covering perhaps only 10 percent of the United States. Not all of the geologic maps at that scale have been converted into GIS-ready datasets. The National Geologic Maps Database, and its associated data viewer, are available via the USGS website and provide search capabilities for users to discover which geologic maps exist and which are available in digital form.

Soils data

Soils are critical for sustaining agriculture and human life. Soils maps have long been recognized for their importance and have been compiled in detail since the mid-1800s by national natural resource agencies, tribal, state, and provincial governments. However, unlike major landforms or even biomes, soils are remarkably complex, varying significantly from place to place and over comparatively short distances by climate, drainage, and bedrock type. As soils are not easily mapped by satellites or sensors aboard airplanes, compiling soils data has always been an expensive and laborious undertaking, which relies extensively on ground truthing and soil sampling.

Esri has created and digitized a global soil raster spatial dataset that illustrates twelve soil types and is based on a reclassification of the Food and Agricultural Organization's (FAO-UNESCO) soil map of the world combined with a soil climate map. It is available

Soils support all life, yet are among the most vulnerable resources, such as this soil in Iowa. Photo by Joseph Kerski.

from the USDA website. A more detailed dataset started in 1986, but not in GIS format, is maintained by the International Soil Reference and Information Centre (ISRIC) as the Global Soil Profile Data. This stemmed originally from the soil profiles, samples, and maps made for the International Soil Museum in 1966 to illustrate the units of the printed FAO-UNESCO Soil Map of the World. The collection has grown to over 1,125 soil profiles from over seventy countries. The ISRIC-WISE Harmonized Global Soil Profile Dataset from the World Soil Information foundation is another example of a global soils database. The dataset is currently available in Microsoft Access format. In Europe, POSTEL is a thematic data center focusing on soil and vegetation from Earth Observation satellite data.

For decades in the United States, the Soil Conservation Service was responsible at the national level for educating and encouraging landowners to adopt best practices in managing soils for agriculture and other land uses. Through a combination of field surveys and supplementary aerial photography, soil surveyors conducted a series of field transects across the landscape to determine the spatial relationships between soil types, terrain, vegetation, and land use. The results were collated into a series of detailed maps, later compiled into county soil books with soils areas reproduced on paper prints of aerial photographs. The expanded vision of the Natural Resources Conservation Service (NRCS), as the Soil Conservation Service became in 1994, included the conversion of these soils books into a digital spatial database. The NRCS now maintains three digital soils datasets (hosted on the USDA website): the National Soil Geographic Database (NATSGO) is the smallest scale data, followed by the State Soil Geographic (STATSGO), and the Soil Survey Geographic (SSURGO), which was collected at about 1:24,000 scale. While it is not intended to be used for crop yield predictions for individual fields, SSURGO does include 18,000 soil series, or groups of soil types, which may be used to identify areas sensitive to development and erosion, and areas best suited to specific crop types. SSURGO data are linked to a Map Unit Interpretations Record (MUIR) attribute database; the Map Unit Identifier (MUID) field links the soil types to their physical and chemical properties. These include pH, depth to bedrock, salinity, most appropriate crops, and water capacity.

Natural hazard data

Natural hazards spatial data include event-related data such as the locations of earthquakes, hurricanes, flooding, and landslides, as well as data indicating which areas are more at risk from those phenomena. As many natural hazards are global phenomena, more global hazard data are available than other data themes. Some of these data began as paper maps created by international relief agencies, national geological surveys, and emergency management organizations. However, depending on the type of hazard, global coverage is still patchy, particularly for less sensational hazards such as coastal erosion and hailstorms.

Global earth events websites from environmental news agencies and other organizations are increasing. Many of these sites contain an RSS feed, a Really Simple Syndication set of data that is updated every time a new natural disaster strikes. However, to use such a feed in a GIS, you need to find a website that offers a special kind of RSS feed with embedded location information—a GeoRSS—such as on Earthpublisher and the USGS near-real-time earthquake and landslide bulletin.

In the United States, several agencies, including the National Oceanic and Atmospheric Administration (NOAA), the USGS, and the Federal Emergency Management Agency (FEMA), provide public domain hazards data. NOAA's Storm Prediction Center provides storms and winds data for download, including hurricanes, significant wind, tornadoes, hail, and blizzards. Volcanic data may be downloaded from Oregon State University's Volcano World website. The Smithsonian Natural Museum of Natural

History Global Volcanism Program, "Volcanoes of the World," is available as an Excel spreadsheet containing latitude and longitude data. A collaborative project between the Smithsonian and the USGS, "This Dynamic Planet," has produced volcano, earthquake, and impact crater spatial data in the form of an interactive web map, but more importantly for the GIS user, as a set of downloadable files. Impact crater information, collated in an impact database, may be obtained in Excel format from researchers Rajmon and Roesler.

While flood data are beginning to become more publicly available as web mapping services, such as the North Carolina Floodplain Mapping Information Service (FMIS), obtaining flood data for ready use in a GIS environment is more difficult. Most flood data in the United States were originally compiled as a set of detailed, paper-based, flood insurance rate maps, not all of which are still available in a GIS-friendly format. The National Flood Hazard Layer (NFHL) is a compilation of these maps but, although the data have been updated, not all areas are complete. The Digital Flood Insurance Rate Map (DFIRM) database is used to create new maps and may be accessed from the FEMA Map Service Center. The Dartmouth Flood Observatory, hosted by New Hampshire's Dartmouth College, maintains a Global Active Archive of Large Flood Events.

Geographic names data

Most atlas editors know that maintaining correct entries for place and feature names is a never-ending task. Names are a critical component to spatial data, as they provide context and history for other base information. The US NGA's Geonet Names Server provides locational information for over four million features with 5.5 million names in its database. Approximately 20,000 features are updated monthly. Toponymic information is based on the Geographic Names Database, containing standard names approved by the US Board on Geographic Names. This increasingly includes native script spellings of some of the names. The site hosts a web map service and also provides responses to individual queries in HTML format. This is easily converted to spreadsheet or text file format and imported into a GIS environment, with coordinates generated from the x,y values.

The EuroGeoNames (EGN) project, funded by the European Commission, aims to address some of the confusion that results when people use different spellings and languages when referring to the same location. EGN, one of the first INSPIRE-compliant data services, provides access to an official geographic names data resource

Place names are a reflection of an area's history, geography, and people. Istanbul is one of the world's true crossroads. Photo by Joseph Kerski.

and includes the official spelling, spellings in other languages, the location of a name, and the pronunciation of a name.

In the United States, the Geographic Names Information System (GNIS) contains every name on every USGS topographic map, as well as others, in its database of over 2,000,000 names. The database assigns a unique, permanent feature identifier; this is the only standard federal key for accessing and integrating feature data from multiple datasets. This reference is essential for data sources such as geographic names, which may have multiple and variable spellings in several different languages and change over time. The GNIS also provides an excellent resource to download and investigate within a GIS as the data include not only the place names and locations of features in the United States and in Antarctica, but also rivers, lakes, valleys, tunnels, schools, and other information.

Street data

Street data have two main purposes and formats, geocoding and transportation. As a geocoding data source, street data are used as an address service, for example, pinpointing the location of a property during an emergency or uniquely identifying the location

Street data, such as this network near Twentynine Palms, California, can be categorized into geocoding and transportation datasets. Photo by Joseph Kerski.

of a home when compiling a census return. This type of spatial street data generally varies a great deal in accuracy, as the prime purpose is simply to geocode a location on the correct side of the street, in the correct block, and in the correct part of the administrative or census area (enumeration district, county, or other area).

Sometimes referred to as postal address data, the format of this type of street data is typically a vector file with address ranges on each side of the street, with a *from* address and a *to* address. Fields include directional prefixes, suffixes, street name, street name type (avenue, boulevard, street, way, court, and other types), and the postal code for that street segment. These street data are the result of considerable investment in time and money to gather and encode the data, which are often collected and maintained by national statistical agencies or local emergency services agencies. In Britain, postcode information is available as a database file, the Postcode Address File (PAF), collected and maintained by the Royal Mail. In 2010, the GeoPlace joint venture was announced, involving local governments in England and Wales and Ordnance Survey, to create a National Address Gazetteer. It has been proposed that GeoPlace would maintain a single address and street database for England and Wales.

The second type of street data focuses on transportation—the type of surface, type of use allowed (pedestrian, bicycle, ski, vehicle), road width, average annual or daily vehicle traffic, and surface maintenance record, among other attributes. These spatial data are typically stored as vector datasets, at increasing levels of detail from the national to the local level. They are sometimes encoded as network datasets, to support the modeling of traffic flow through the transportation system.

With the exception of the DCW and the Global GIS Data CD series, two data sources mentioned before, very little global public domain street or transportation data have been available. Users had little option other than to pay for larger-scale national street and transportation datasets. These were produced either by national mapping agencies—in Britain, Ordnance Survey developed the Integrated Transport Network (ITN) Layer as part of the OS MasterMap product—or private mapping companies such as DeLorme, Garmin, and Tele Atlas, which developed their mapping products in response to the increased demand for vehicle routing data.

As a result, there is no shortage at present of easily-available, GIS-ready, street and transportation spatial data. Not only are multiple themes and scales available, but even *within* a single theme, there may be several data options to choose from. This presents opportunities and challenges for the data user. Say that you want to use some street data as a backdrop for your demographic analysis; the private companies providing vehicle routing information supply excellent quality street data, which offer certain advantages over public domain data. For example, the data are up-to-date and made available in many different formats with multiple attributes, including address ranges. Obtaining data from value-added data resellers can save a lot of time and effort, but if you really want to use public domain data because of limited funds or to avoid restrictions on end use, you still have some options to consider.

Recent years have seen the emergence of the OpenStreetMap (OSM) project, an initiative to create a global street map that is free for all to use and share. The data are supplied by registered volunteers conducting ground surveys and therefore have the additional advantage of potentially being the most up-to-date street data on the planet. We will return to this project later in the book when we discuss volunteered geographic information (VGI).

In the next section, you will compare two other public domain street centerline datasets and evaluate which would be better for certain GIS applications. Perhaps more important than the specific datasets themselves, however, is the decision-making process involved in selecting the right dataset for your application.

Comparing two public domain transportation layers

If you wish to use public domain street data for the United States, two of the most common options are DLGs or US Census Bureau TIGER/Line (Topologically Integrated Geographic Encoding and Referencing) files. These data sources were created for two different purposes by two different federal agencies. Both involved millions of dollars of investment by federal agencies, thousands of person-hours, and decades of work. DLGs were created in the early 1980s by the USGS, as digital representations of its 1:100,000-scale topographic maps. As the topographic maps followed US National Map Accuracy Standards, the locational accuracy of the resulting digital data were very high. However, most streets were never intended to be named on topographic maps, and as a result, the DLG streets layer contains few names. The exceptions are some federal and inter-state highway designations. Conversely, TIGER files from the US Census Bureau were created specifically for the decennial censuses of 1970 and 1980 to address reference census forms in urban areas. They represent an extension of earlier digital streets data from the Geographic Base File/Dual Independent Map Encoding (GBF/DIME) files. GBF/DIME files, extended to cover rural areas for 1990, became the first national street network. As it provided street and address ranges, it was the starting point for Tele Atlas' street data. It was also adopted for various web mapping services, including MapQuest, Yahoo! Maps, and Google Maps.

The streets in urban areas were originally hand drawn by census enumerators, working without the benefit of GPS or aerial photos. With the main objective of referencing census responses to the correct block and therefore to the correct block group, county, and state, an accurate spatial representation of the street pattern enclosing that block was of secondary importance. As rural areas were completed with DLGs from the USGS during the 1980s, and later during the 1990s and 2000s, with the Bureau of Census' own TIGER Update Program, the spatial accuracy of the TIGER dataset has steadily improved. Even though they both include a streets layer, DLG and TIGER data remain quite separate and distinctive public domain spatial datasets, as the following example illustrates.

	DLG	TIGER
Producing Agency	US Geological Survey	US Census Bureau
Collection Scale	1:24,000, 1:100,000, and 1:2,000,000	1:100,000
Spatial Accuracy	Excellent; produced from aerial photographs using National Map Accuracy Standards	Mixed: Originally poor in urban areas where original data contains remnants from 1970 and 1980 GBF/DIME files, but georectification has improved accuracy. Good in rural areas where data came from DLG. Urban and rural accuracy improving due to TIGER update program.
Currency	Rapidly becoming outdated. Much was collected from 1975 to 1995 but may be based on topographic maps from the 1950s or 1960s.	From the 2000 Decennial Census (and updating to 2010), but TIGER update program in selected counties is improving currency between Decennial Census years.
Format	Cumbersome to read into a GIS; standard and optional formats are text-based (ASCII), and require conversion from their native format to something that a GIS can import.	Cumbersome to read into a GIS in its native format, but most TIGER files have been converted into easily-read GIS-ready files.
Attributes	Cumbersome to manipulate in a GIS due to its numeric format, and due to major-minor code pairs (for example, 170 403) in multiple columns that are not in the same order from file to file or even from feature to feature.	Easier to manipulate in a GIS; A single alphanumeric Census Feature Class Code that represents the type of road, for example: A43.
Main Utility	Spatially accurate base data	Street data can be address geocoded; other layers can be linked to Census demographic information and thematically mapped.
Access	http://www.usgs.gov/, accessed 01/17/2011 - search for 'geographic data download' and 'earth explorer'	www.census.gov, www.Esri.com, www.gisdatadepot.com, and others.
Distribution Cell Size	By 7.5-minute by 7.5-minute cell at 1:24,000-scale, by 30-minute by 1-degree cell at 1:100,000-scale, and by state by 1:2,000,000-scale.	By county and county equivalent (independent cities, boroughs, and parishes).
Content	Streets, hydrography, vegetation, non-vegetation (sand, lava), contours, survey markers; up to 11 layers at 1:24,000-scale, but most geographic areas only have a few layers compiled.	Streets, hydrography, landmarks, statistical area boundaries; political area boundaries, and more.
Vertical Integration	Layers not vertically integrated with each other.	Layers are vertically integrated.
Horizontal Integration	Some layers not horizontally integrated with adjacent topographic sheets.	Layers are horizontally integrated.
Layer coverage	Only selected layers have been compiled at the 1:24,000-scale; most layers have not been compiled.	All layers have been compiled for entire country.
Spatial Coverage	The United States at 1:100,000 scale and 1:2,000,000 scale	United States at 1:100,000 scale.

Comparing digital line graph and TIGER data formats. Data courtesy of US Census Bureau.

DLG VS. TIGER

Based on the preceding information, name two applications where using DLGs would be more appropriate.

Name two applications where using TIGER files would be more appropriate.

Why did you name the appropriate applications that you did?

Demographic data represent some of the oldest data that have been linked to location. Photo by Joseph Kerski.

Demographic data

Governments have been collecting data, in some cases for hundreds or even thousands of years, on population, housing, agriculture, economics, and other demographic variables. The earliest documented censuses date from those undertaken 500–499 BC in the Persian Empire. Sweden's 1749 census was, however, the first modern census. Many countries collect data at several geographic levels, ranging from the country as a whole, to states and provinces, regions, cities, districts and individual city blocks. Some conduct a recurring census every ten years or at other intervals, and some use the same or similar statistical areas as in previous census years. This has the additional benefit of allowing GIS practitioners to conduct historical comparisons. The number of questions asked, and level of detail collected varies considerably among the statistics agencies conducting the survey. The quality of data provided in the returns also varies, due to the mobility of the population, the manner in which the data are collected, the type of government that is collecting the data, and many other factors. As a result, some sections of the population may be underrepresented.

Most census data online tends to be either at the coarsest level of geography, or via web-query forms that require a great amount of reformatting to make the data GIS-ready. Occasionally, data from statistics agencies are available in spreadsheet format. Less frequently, statistical boundaries are available; these are perhaps the most useful format for a GIS analyst, as they provide a means to map the accompanying statistical data. These areas are sometimes available from national statistics agencies, but typically are more difficult to obtain than the statistics themselves. Some data from Statistics New Zealand and the UK National Statistics agency are available from their respective websites. Some historical census data are also available, for example, for the United Kingdom back to 1801 and for the United States back to 1790, via the National Historical GIS project from the University of Minnesota.

In this example, Census Tract 011503, outlined in yellow, is divided into two block groups, 011503–1 in the south, and 011503–2 in the north. Each block group is divided into a dozen or more individual city blocks. Data courtesy of US Census Bureau.

The US Census Bureau collects data for political areas and statistical areas. The political areas include American Indian reservations, cities, counties, and states. Statistical areas, on the other hand, are not political entities but are strictly for reporting purposes. These include, but are not limited to, census tracts, block groups, census-designated places, and blocks. Blocks in a city or town are bounded on all sides by streets. Blocks in a rural area may be much larger, and may be bounded by roads, streams, or railroads. Groups of blocks are called block groups, groups of block groups are called census tracts, and census tracts nest within counties or county equivalents, such as parishes, independent cities, or boroughs. The relationship between the various statistical areas is shown in the following example.

To solve problems with public domain census data, a typical workflow is to download the statistical areas that you wish to work with. Next, download the statistical data, such as educational attainment, income, median age, owner-occupied versus renter-occupied units, commuting patterns, family structure, ethnicity, or housing type. The most important thing is to ensure that your statistical data cover the same level of geography (such as block group or enumeration district) as the statistical area features you are working with. The statistical areas form the physical boundaries of your area of investigation (the *G* part of GIS), while the statistical data provide the facts and figures for that area (the *I* part of GIS). If you are working with block group areas, for example, make sure that the demographic data you use are at the block group level.

To use census data in ArcGIS, the areas must be loaded into a geodatabase or maintained in a shapefile. The associated statistical data should be stored in a spreadsheet or delimited text file. To visualize and analyze the data, the statistical data must be joined to the area spatial data layer based on a common field.

British statesman and naturalist John Lubbock once said, "What we see depends mainly on what we look for." How a problem is represented can greatly influence how easily that problem is analyzed and solved. The quality of the data used, their accuracy and the level of detail affects every task and analytical procedure run on those data. This, in turn, may have a profound impact on the results of the analysis, often with far-reaching implications. For many GIS projects, data quality can be an area of great uncertainty.

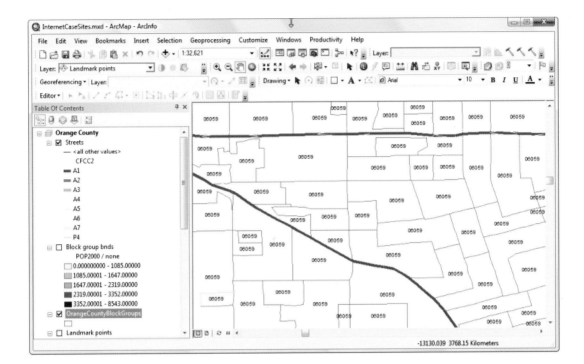

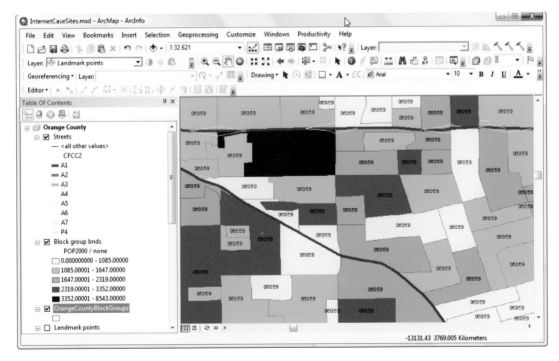

Block group boundaries without demographic data, on the top, and with demographic data, mapped by total population in graduated color, on the bottom. Data courtesy of US Census Bureau.

Data quality: "There Be Monsters!"

Tackling any spatial problem using public domain data requires an understanding the characteristics of data at every step in the entire process, including the origins of the data, why each layer was created, who created the data, how the data were created, and the precision and accuracy of the data. For centuries the printed word was treasured and its veracity often unquestioned, perhaps because creating and contributing to these documents were beyond the means of the masses. Today, anyone can easily write on any topic and place it in view of millions of people through blogs, wikis, tweets, RSS feeds, and a multitude of networking and discussion outlets. However, just because it is so easy for virtually everyone to share their thoughts doesn't necessarily mean that every published word is factually correct. Similarly, just because maps today are produced in digital form, can be updated almost instantly, and are available just about everywhere—via the Internet, on a handheld computer, or a phone—does not mean they are error-free. Nothing could be further from the truth. All maps are inherently inaccurate because they attempt to represent the three-dimensional earth on a two-dimensional piece of paper or computer screen. Map projections attempt to minimize the distortions in the distances, direction, shapes, angles, or sizes, but some distortions will always remain; there is no perfect map. The intended use of maps may also influence their accuracy; some maps are scientific documents, others are produced for illustration only. The standards adopted to produce each map vary widely depending on the date, the producer, the purpose, and the scale.

Digital maps are often no more accurate than the original source paper documents. In many cases, legibility was often deemed more important than the spatial accuracy. For example, roads on USGS topographic maps running parallel to railroads were offset from the railroads so that both could be legible to the map reader. In Britain, some contour data were omitted to make the original Ordnance Survey maps more legible. In addition, where they are coincident with manmade features, contours may be broken, and are not shown at all in active quarries, gravel pits, spoil heaps, or open cast mines. In other words, USGS, Ordnance Survey, and many other topographic maps were originally created as *cartographic* products, rather than representing the location of every feature as *geographic* products. In their corresponding digital datasets, roads were often not relocated to their true location and contours were not reinstated where they had been omitted. Anyone using the spatial datasets derived from these maps should bear these facts in mind. Typically, this type of information is not included in the metadata.

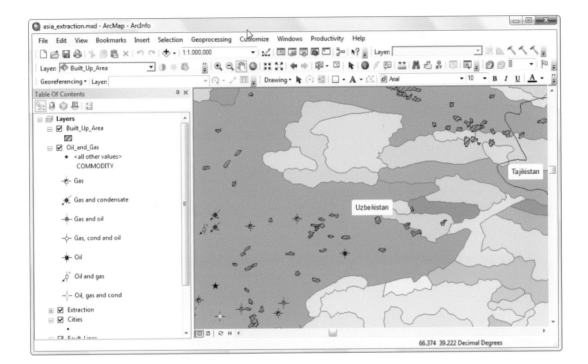

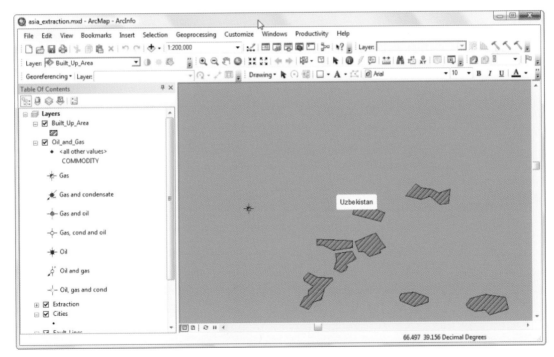

In the top image, oil, gas, and geology for central Asia examined at the same scale at which it was created (for reference purposes)—1:1,000,000. In the bottom image, the data are being examined at 1:200,000-scale. How confident can you be in making decisions at 1:200,000-scale when the data are no more accurate at this scale than they were at 1:1,000,000?

Data courtesy of US Geological Survey.

GIS HAS ROOTS IN DIAL-BEFORE-YOU-DIG INITIATIVES

One of the many driving forces behind the emergence of digital spatial record keeping came from pipeline incidents. In 1970, seventy-nine people were killed, 380 injured, and 101 houses were destroyed by fire when a hapless construction worker in Osaka, Japan, inadvertently punctured a gas pipeline. Rapid urban development had resulted in the chaotic arrangement of pipelines and, as the positions of the pipes were largely undocumented, neither the worker nor the construction company had access to detailed plans of the new network. Utility companies were subsequently compelled to make plans of their networks and make those data available to others (Kubo 1987).

Furthermore, the capability of a GIS to examine the earth at an infinite variety of scales often creates the illusion that data are more accurate than they really are. In the following examples, public domain energy and geologic data were originally created at 1:1,000,000 scale but are being viewed at 1:20,000 scale in a GIS.

Why should you care about scale when the tools are available to allow you to zoom, pan, and examine the data at an infinite variety of scales with consummate ease? A dataset is no more accurate at a scale larger than the scale at which it was created. Whether a gas pipeline is 20 meters or 40 meters from a road may not be important to a researcher studying energy networks in Europe but that difference may be critical for the gas company or municipality sending out maintenance crews to inspect and maintain the pipeline infrastructure. Both groups may regard the data as fit-for-use and as sufficient quality but they clearly have different requirements. At best, incorrectly represented facilities may result in the energy company having to incur additional expenses to procure extra equipment. At worst, if field crews start digging in the wrong place, injuries and fatalities may result.

But what is quality and how do you measure it? Even the Chartered Quality Institute, as the self-professed "only chartered professional body dedicated entirely" to the subject, admits that quality is difficult to define. Nevertheless, it defines quality in terms of innovation (in product, service, management, and processes) and care (to society, the environment, and people). Quality encompasses more than what meets the eye, and has as much to do with its impact on the user and the organization as on the object or problem being addressed.

ISO, in Standard 8402 (1994), defined quality as "the totality of characteristics of an entity that bear on its ability to satisfy stated and implied need." The quality of any product or service cannot be established in a vacuum and depends on the requirements of the user. If the characteristics of the product or service meet the user's requirements, the product or service is deemed to have high or excellent quality. If the characteristics do not meet the requirements, the product or service is considered to be of low or poor quality.

Just as one person's choice of music may be dreadful noise to someone else, quality is a subjective concept, dependent on the recipient's perception and requirements. Quality may also span a range, rather than simply "of quality" or "not of quality." Crosby (1995) defined quality as "conformance to requirements." Each characteristic of quality is tied to a requirement and is an inherent feature or property of a product, process, or system. A requirement is an expectation, a need, or an obligation stated by the project managers or the data users.

So, instead of asking, "is the spatial data of good quality?" you should ask "is the data of sufficient quality for my project?" In other words, rather than asking "is it good?" ask "is it good enough?" While this may sound like an abrogation of duty, it places the burden of responsibility where it belongs—with the user. It is the data user's responsibility to determine if the data are fit to use for his or her particular application.

As we have already seen, using any data involves uncertainty and it has potential to affect the decisions that are made based on the data, as well as the organizations themselves, particularly where administrative decisions are subject to judicial review. While many consumers would not consider purchasing a computer without an instruction manual and a warranty, organizations still spend a great deal of money purchasing spatial data with incomplete, or even worse, no supporting documentation. If the analysis and presentation of geographic information is to be recognized as a valid field of scientific endeavor, then you as a GIS user must be able to describe how accurately the information you base your decisions on represents the truth. You must learn how to recognize the inherent uncertainty in the data you use, to minimize that uncertainty, manage it, and communicate it in your application. When you begin to use any dataset, you must ask if each decision you reach based on those data involves low or high risk to you, your organization, and any others who consume the information products you produce. Ideally, you should know what the final product quality specifications are before beginning a project; in practice, you may have to assess it later, once the level of uncertainty has been quantified. Decide what product quality will be acceptable for your GIS application and whether the uncertainty present is appropriate for the tasks involved. You may reject the dataset as unsuitable, choose a different dataset, or generate the dataset yourself. Or, you could accept the dataset, acknowledge the uncertainty, and change the parameters of your study.

While it is the user's responsibility to decide fitness for use, it is the data producer's responsibility to provide "truth in labeling." Data producers should provide information about who created the data, when the data were created, for what purpose, the scale the data were created at, the source documents used, the content, field names in the tables, the map projection, and other critical information. This is generally provided in the form of metadata, the information that describes the data. Metadata contain information about the dataset's content and quality. Traditional hard copy maps, the source of much digital data, contained accuracy statements such as reliability diagrams and estimates of positional error. Although far from perfect, these statements were often lost when the data were converted to digital format, either because they were in the margin, on the back of the map sheet, or elsewhere. In addition, as a GIS may represent values and absolute locations with an almost limitless number of digits, such as 43.10474 north latitude, people may believe the digital data are more accurate than a paper map. Data in a GIS may be more *precise*, but not necessarily more *accurate*.

Unfortunately, metadata are often missing either because they were never encoded in the first place or have been lost in transfer between users. If metadata are missing, determining fitness for use becomes even more of a challenge.

Spatial data users think they want the most accurate data available. The reality is that few users would be willing to either wait until such data have become available, or pay the high costs involved in producing accurate data. Few private data providers could sustain their operations solely by providing only top quality data; few public data providers would receive the necessary funding from their national, state, or local governments to produce such data. In either case, regardless of who was producing the data, data users would ultimately suffer. Not only would the cost be prohibitive, but very high quality data would require more resources (computer memory, storage space, and so on) and would be unwieldy to manipulate on a local computer or server.

Although these quality issues have existed ever since cartographers began drawing paper maps, they affect all types of data. Failing to appreciate the quality of data with regard to maps is particularly dangerous for several reasons. First, the quality of map data is more easily hidden than with something more obvious such as food, a hotel, or car maintenance. Second, maps are powerful tools used for decision making on a daily basis and as such, any misinterpretation of the information may have potentially serious and lingering consequences. Third, a GIS-based project typically uses a number of map layers from a variety of sources, each created for different needs, at different scales, and transformed from different projections. These issues should not detract from the power of GIS analysis, but are among the core issues that users must continually keep in mind. Never assume the data are of sufficient quality for your use. Be critical of the source and always examine the metadata file.

There are five measures of accuracy you should consider when determining whether the spatial data are fit for your use. These include:

Positional accuracy: How close are the locations of the objects to their corresponding true locations in the real world? Will this be sufficient for your needs?

Attribute accuracy: How close are the attributes to their true values?

Logical consistency: If you used this dataset with other datasets, will its spatial characteristics and its attributes cause strange juxtapositions or illogical associations, such as a road that is also a canal or a building height that is below the elevation of the land surface? Does every area have a label point? Is the dataset consistent with its definitions?

Completeness: Does your proposed dataset completely capture all of the features that you need?

Lineage: Does the lineage of the data—who created the data, how the data were created, what methods were used to collect the data—meet your requirements?

READ AND RESPOND

Read Jeff Thurston's "Perspectives" column about data quality, "Can the geospatial industry advance much further without addressing quality seriously?" (*V1 Magazine*, 2009).

On which issues does Thurston believe the geospatial community is spending too much time? Where is the community spending too little time?

Why do you think he believes that attention to quality is tied to the advancement of the field of GIS?

What forms the link between a "robotic future" and spatial data quality?

Do you agree with Thurston's position?

Remember that accuracy is not the same as precision. A dataset's accuracy is the closeness of results, computations or estimates to true values, or values accepted to be true. A dataset's precision, on the other hand, is the number of significant digits (decimal places) in a measurement. Are your data precise and accurate enough for your needs? If not, can you live with your data? More importantly, can your end-users live with the products or decisions that are derived from the data you used?

Census data quality

A census is always challenging in terms of collecting, tabulating, and disseminating information. While improving technologies in all three areas have made national statistics more readily accessible and more easily analyzed within a GIS than ever, it is still important to understand how and why those data were collected. In addition, due to the expense of conducting a national census, the data are increasingly supplemented by projections and sample surveys. For example, in the United States, a long form of the decennial census (every ten years) questionnaire was previously given to one of every six households. This was replaced after the 2000 Census for cost and efficiency reasons with the American Community Survey (ACS). Comparing ACS data to decennial census data is a bit like comparing apples to oranges. The ACS is a statistical representation of the population but not directly comparable to the decennial census's complete count of the country. But is the complete count really complete? While mail-out-mail-back questionnaires have been used for decades and help increase participation, by not relying on a census taker knocking on every door, certain segments of the population are often underrepresented, such as the homeless, college-age students, mobile urbanites, and some ethnic groups, particularly those living in non-English speaking households. During the 2010 Census, many people under thirty years old were not responding as expected. These representatives of the digital generation tended to throw away their census questionnaires, as they did most other paper items received in the mail. Yet, while a dynamic population makes for a greater challenge for enumeration, it's one of the main reasons for conducting a census in the first place. Furthermore, if costs continue to rise and people continue to express concerns about being surveyed, governments may rely on their own internal records of people to conduct a census rather than actively survey the population. This would raise new concerns, not only about data quality but also of privacy.

Census questions themselves may also be misinterpreted and even fundamental questions on age, race, and place of residence have changed over the years. Questions on race in particular have proven controversial and resulted in protest from various groups who feel they are not adequately represented. While the Census Bureau holds hearings every decade to establish the appropriate definitions and required input from the population, questionnaires remain self-enumerating, completed by the ambiguous head of household, and subject to individual interpretation. Responses are, for the most part, not verified.

Lastly, the US Census is conducted, first and foremost, to reapportion the US House of Representatives based on the population of the states. Understanding why and how it is undertaken is the first step towards appreciating the nature of census data. Despite shortcomings and limitations, census data have become an invaluable data resource for funding and managing thousands of programs at local and national levels.

Summary and looking ahead

In this chapter, we have considered some of the different types of vector data that are available. We have explored some core global and country-specific vector public domain spatial datasets, and reviewed some of the sources for obtaining those datasets. We also discussed spatial data quality and its implications on analysis within a GIS environment.

In the next chapter, we will use additional datasets from other national agencies. We will do this in conjunction with an exercise to locate a wildfire monitoring tower in the Loess Hills of eastern Nebraska.

References

Burrough, Peter A., and Rachael A. McDonnell. 1998. *Principles of Geographic Information Systems.* 2nd ed. New York: Oxford University Press: 356.

"Can the Geospatial Industry Advance Much Further Without Addressing Quality Seriously?" 2009. *V1 Magazine* (August 14). http://www.vector1media.com/vectorone/?p=3431.

Crosby, Philip. 1995. *Philip Crosby's Reflections on Quality.* New York: McGraw-Hill: 144.

Kubo, Sachio. 1987. "The Development of Geographical Information Systems in Japan." *International Journal of Geographical Information Science, 1362-3087,* 1(3):243–252.

Lehner, Bernhard, Kristine Verdin, and Andy Jarvis. 2008. "New Global Hydrography Derived from Spaceborne Elevation Data," *EOS Transactions.* American Geophysical Union, 89 (March) (10): 93–94.

Olson, David M., Eric Dinerstein, Eric D. Wikramanayake, Neil D. Burgess, George V.N. Powell, Emma C. Underwood, Jennifer A. D'Amico, Illanga Itoua, Holly E. Strand, John

C. Morrison, Colby J. Loucks, Thomas F. Allnutt, Taylor H. Ricketts, Yumiko Kura, John F. Lamoreux, Wesley W.Wettengel, Prashant Hedao, and Kenneth R. Kassem. 2001. "Terrestrial Ecoregions of the World: A New Map of Life on Earth." *BioScience* 51 (November) (11): 933–938.

RASTER DATA AND PRIVACY ISSUES

Introduction

In chapter 2, you examined spatial data models, explored major types of vector data, and studied some issues surrounding spatial data quality. You located, downloaded, formatted, and analyzed vector data to select the best sites for a business in Orange County, California. In chapter 3, you will examine raster datasets with the following questions:

Who produces these datasets?
Which are in the public domain?
What themes do they include?
How can they help the decision-making process?

Raster data are now available in more wavelengths and formats, with geo-tag features, and at a higher spatial resolution than ever before. You can now take a georeferenced photograph of a person in a particular location and upload it to the Internet. This chapter also will discuss privacy concerns and how they may be reconciled with public access to spatial data.

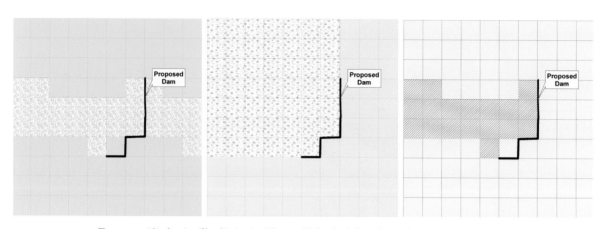

The raster grid indicating if land is riparian [1] or not [0] (on the left) can be easily combined using a multiplication overlay with land that would be flooded [1] or not [0] by a proposed dam (in the center), resulting in a binary raster layer indicating which riparian zones would be flooded [1] (at the right).

Advantages to raster data analysis

Raster data offer a powerful yet simple format for advanced spatial and statistical analysis, which support both surface analysis and rapid overlays, and provide a uniform way to store information. In the following example, you have a two-category dataset indicating whether land is in a riparian zone and you had another two-category dataset of the same area indicating whether land would be inundated by a proposed dam to illustrate which riparian zones faced flooding.

Sometimes you don't have a choice between storing your spatial data in a vector or raster format. For example, an aerial photograph or satellite image must be stored in raster format. However, many other features and phenomena, ranging from seasonal changes in the Arctic Ocean ice cap to crime patterns in a city, may be modeled as either vector or raster. Features are represented as collections of cells in the raster model, losing their unique identities and specific boundaries. So, as a good general rule, the raster data format works best when the main focus is on the spatial relationships of the phenomena represented by the geographic features, not the features themselves. For example, analyzing the distribution of aridity versus land cover can reveal underlying processes; the names of the clusters of arid lands and the ecoregions containing certain land cover may be irrelevant to that investigation. Vector data storage is best suited to situations in which the features themselves are of equal concern as the phenomena they represent.

There is, however, no single rule dictating whether you must use the vector or raster model. The more informed you are about your data and their limitations, the more likely you will best decide how to model your data. Making the right choice at this stage will positively influence what you can achieve with your GIS-based analysis.

Recall from the discussion in chapter 2 that the raster data model is actually a very simple structure comprising a regular grid of cells arranged in an array of rows and columns. However, like vector spatial data, raster data come in a wide variety of formats, cover a wide variety of themes, and are available at a variety of scales. To better understand and use raster data effectively in a problem solving environment, we will review the relative merits of raster data and their two main types in the next section.

Categories of raster data

Raster data can be thought of as either discrete or continuous. Discrete raster data are also known as thematic data, as they represent one specific theme describing the earth's surface. They are also referred to as categorical data, as they cover one specific category. These themes, or categories, could be wetlands, land use, soil type, or any other discrete phenomena that can be represented in this way. Discrete rasters most often represent objects that may also be represented as vector data. As these objects have clear, definable boundaries, why, then, are they often represented as rasters?

Sometimes it is more efficient to store these datasets as raster data. Consider land use, for example. Land use data have a definable set of categories, such as pasture, industrial, or low-density residential. However, as land use varies so much across the surface of the earth, a vector area land use dataset, even for a small area mapped in moderate detail, may easily include millions of areas. Using such a complex dataset to solve a particular spatial problem may result in longer data-processing times and reduce the overall efficiency of the decision-making process. A raster overlay, on the other hand, may be less complicated and easier to process, while still providing the necessary level of detail.

Another reason why these datasets are often best stored as rasters is accuracy. In the forest lands in northern Wisconsin, deciduous trees in the south give way to coniferous trees farther north in the boreal forest. Rather than representing the forest as areas that are either deciduous or coniferous, which implies an identifiable "line" or boundary separating the two, the landscape may be more appropriately represented as grid cells that include a percentage of deciduous tree cover, ranging from 90 percent in the southern part of the study area to 10 percent in the northern part. This model reflects the natural, more gradual changes in forest cover that occur as climatic and soil conditions change from south to north.

Continuous data, by contrast, often represent phenomena that are under constant change, such as movement of wildfires, air, or ocean currents. Changes like this are efficiently modeled using the raster format. Phenomena in which each location is a measure of its relationship from a fixed point, such as elevation, slope, or aspect, are also best represented as continuous, or surface, rasters given that data values are stored as floating point numbers rather than integers. As such, there are an infinite number of potential data values for each cell. For example, the elevation above sea level of each cell in a particular region, whether represented in meters or in feet, could be any number. Each elevation may be different from that of any other cell's elevation, particularly if the number of significant digits (decimal places) is high.

Another way to classify raster data is according to band. Some rasters have a single band, or layer, while other rasters are comprised of multiple bands. A band is represented by a single matrix, or grid, of cell values. The term band refers to the color band on the electromagnetic spectrum. A digital elevation model (DEM), which we will discuss in more detail later in this chapter, contains one—and only one—value for a cell that represents the elevation of the land surface in that cell. A panchromatic orthophoto, a single-band aerial photograph that has been modified for use in a GIS, is another example.

A raster with multiple bands contains multiple matrices of cell values that represent the same spatial area. The most common examples of multi-band rasters are satellite images derived from sensors sensitive to different parts of the electromagnetic spectrum. A true color orthophoto collects data that represent red, green, and blue light. Landsat imagery, which we will also examine later, is also a multi-band raster dataset as its sensor collects data in seven different bands, including the visible, near-infrared, and mid-infrared regions of the spectrum.

Raster data uses

Using raster data in the GIS decision-making process typically involves one of a number of possible applications for the data. Perhaps the oldest and most common use for raster data is as a base layer behind other vector or even other raster datasets. The basemap provides the data analyst with a context and reference point, for example, real world entities (such as buildings and coastlines) depicted in an aerial photograph or satellite image. In addition to photographs and images, scanned historical, topographic, street, or thematic maps are also used as basemaps.

Raster data are used as a surface map, most commonly as an elevation or topographic map, but could just as easily illustrate population density or snowfall. Rasters are also used as thematic map layers, representing such phenomena as soil type, zoning, or land use.

These three uses for raster data have been common practice since the 1980s, but more recently, rasters have increasingly been used as attributes of a feature. These attributes may include photographs taken on the ground, scanned from original prints, or taken with a digital camera. They may also include scanned property deeds, outline sketches drawn at that location, and digital graphs or charts of the height of a stream gage. As an attribute of a geographic feature, these rasters are often stored in the same database for ease of access and improved data integration.

One specialized category of raster data, remotely sensed imagery, is so named because the data are recorded from a distance, without physically sampling the phenomenon under study. These data include aerial photographs and satellite images. Remotely-sensed images are invaluable for a number of reasons. First, they capture data for large areas of terrain in a uniform, continuous manner much more cheaply than mapping the same areas using ground surveys. Second, remote sensing cameras and sensors can detect light from wavelengths that are undetectable to the human eye, extending the types of features or phenomena that may be mapped. Third, when an image is taken, features are recorded at a specific moment in time. Images of the same location taken at different times can therefore be used as historical snapshots to assess how conditions vary during a day, a season, or across many years or decades. Fourth, remotely-sensed data can be converted to spatial data that have a high degree of geometric accuracy. This is due to the development and refinement over many decades of post-processing methods which remove distortion from aerial photographs and satellite images. As a result, remotely-sensed raster data have become the source of a great deal of base spatial data in both vector and raster formats.

Raster data quality

Effective decision making based on spatial data requires an understanding of the data, not simply the way the data are represented. Recalling our discussion in chapter 2 about data quality, does quality matter in the raster world? Absolutely. You can zoom in on a raster image so closely that individual cells can be seen. These cells, or pixels, represent features or phenomena in the raster data model and the resolution of cells is the chief measure of data quality.

Spatial resolution

A significant component of raster data quality is spatial resolution. A dataset with 1-meter spatial resolution is more detailed than a dataset with 30-meter spatial resolution. A higher spatial resolution means that there are more pixels per unit area. The spatial resolution of passive sensors depends primarily on what is known as the instantaneous field of view, or IFOV.

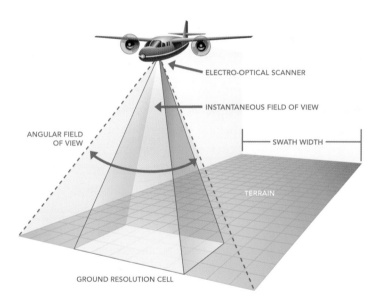

The instantaneous field of view (IFOV) from the sensor to the ground. Source: Esri.

The IFOV is the angular cone of visibility extending from the sensor to the ground; this determines the area on the earth's surface that can be seen from a given altitude at one moment in time. The size of the area viewed is determined by multiplying the IFOV by the distance from the ground to the sensor. The area on the ground is the resolution cell and determines the spatial resolution. If a feature on the ground is greater in size

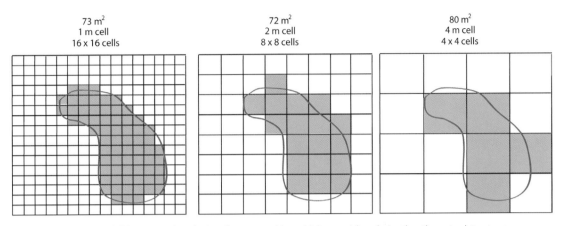

73 m²
1 m cell
16 x 16 cells

72 m²
2 m cell
8 x 8 cells

80 m²
4 m cell
4 x 4 cells

The raster dataset on the left has a spatial resolution of 1 meter and is at a higher spatial resolution than the raster dataset on the right, which has a spatial resolution of 4 meters. Note that the size of each cell is 1 square meter on the left and 16 square meters on the right. As the area depicted in orange changes depending on the cell size (73 square meters on the left and 80 square meters on the right), the spatial resolution affects what the GIS user measures as the areal extent of features and, therefore, any analysis that considers those features.

than the resolution cell, then it can be detected. Conversely, if it is smaller, it cannot be detected—a pond encompassing 10 square meters would not be detectable from a sensor with a 30-square-meter spatial resolution.

It is assumed that point data represented as rasters occupy the entire area of the cell, so the smaller the cell size, the closer the cell should be to the location in the real world. However, no matter what the cell size is, some generalization occurs. This is especially apparent with lines and areas represented in the raster data model. Lines are represented as chains of spatially connected cells, either adjoining on one or more sides, or on their corners. If a line dataset is converted into a raster dataset having a resolution of 1 square kilometer, then a road, for example, would be represented as being 1 kilometer wide, which obviously does not represent its width in the real world. Rasters representing areas are simply a group of contiguous cells with the same value. Recall from our vector data model discussion that small-scale data can be examined at 1:1,000 scale, but that does not mean that the data are accurate to that scale. Similarly, data in raster format can be presented at very fine resolutions, but that does not mean that the data have been collected at that resolution. So, as with vector data, knowing the source and understanding the limitations of raster data is paramount.

Spectral resolution

There is more to resolution than spatial resolution alone. Spectral resolution has to do with the number of spectral bands a dataset has and, therefore, is only applicable to satellite and aerial image rasters. Spectral resolution relates to the ability of a sensor on a satellite or aircraft to distinguish between wavelength intervals in the electromagnetic spectrum. A higher spectral resolution corresponds to a narrower wavelength range for a particular band. A grayscale orthophoto has a low spectral resolution because it captures data from a wide portion of the spectrum—indeed, across all of the visible range—whereas hyperspectral sensors collect data from up to hundreds of very narrow spectral bands. Water and vegetation may be separated by using broad wavelength ranges, but different rock types require a higher spectral resolution to distinguish between them. The hyperspectral approach is able to distinguish between the small variations in the spectral signature of phenomena by comparing it to the spectral signature of known objects or materials.

Temporal resolution

Temporal resolution refers to the frequency with which images are captured over the same place on the earth's surface and, therefore, again only applies to satellite and aerial image rasters. It is sometimes referred to as the *revisit period*. A sensor on a satellite that captures data once every three days, therefore, has a higher temporal resolution than the multispectral scanners on the first Landsat satellites, which could only revisit the same place on the earth's surface every eighteen days. The temporal resolution depends on the satellite's latitude, its capabilities (such as whether it can change the angle to which it is pointing) and the amount of overlap in the image swath (the amount of ground covered lengthwise by the sensor).

Radiometric resolution

Finally, radiometric resolution represents the ability of a sensor to distinguish objects in the same part of the electromagnetic spectrum or, put another way, it is the ability to discriminate between very slight differences in reflected or emitted energy. Image data are generally displayed in a range of gray tones, with black representing a digital number of 0 and white representing the maximum value. Image data are represented by positive numbers from 0 to one less than a selected power of two. This corresponds to the number of bits used for coding numbers in a computer. All numbers in a computer are stored as binary numbers—0s and 1s. The number of levels available depends on how many computer bits are used in representing the energy recorded. If a sensor used 8 bits to record the data, there would be 2^8 or 256 available values from 0 to 255. However, if only 4 bits were used, then only 2^4 or 16 values would be available with less radiometric resolution. A Landsat band is typically 8-bit data, while a GeoEye IKONOS band is typically 11-bit data. Therefore, the IKONOS band has a higher radiometric resolution.

Choosing the appropriate resolution

How do you select an appropriate spatial resolution? One guideline is to choose a resolution that is a factor of ten times finer than the size of the features you need to identify. For example, if you want to visually delineate features with a minimum size (also referred to as the *minimum mapping unit*) of 1 kilometer (1 kilometer × 1 kilometer), a 100-meter spatial resolution is probably sufficient. However, if you need to identify tree crowns that are 3 meters by 3 meters in size, you would probably want to select a 1 meter or finer resolution.

A mathematical relationship exists between the map scale and the image resolution.

"The rule is: divide the denominator of the map scale by 1,000 to get the detectable size in meters. The resolution is one half of this amount. Of course the cartographer fudges. He makes things which are too small to detect much larger on the map because of their importance. But this cannot be done for everything so that most features less than resolution size get left off the map. This is why the spatial resolution is so critical." (Tobler 1987 and 1988).

With these guidelines in mind, the following illustration suggests the appropriate spatial resolution for a selected set of map scales.

Map Scale	Detectable Size (in meters)	Spatial Resolution (in meters)
1:1,000	1	0.5
1:5,000	5	2.5
1:10,000	10	5
1:50,000	50	25
1:100,000	100	50
1:250,000	250	125
1:500,000	500	250
1:1,000,000	1,000	500

The appropriate spatial resolution for a selected set of map scales.

The spatial resolution needed to detect features at a map scale of 1:50,000 is approximately 25-meter [50,000/ (1,000*2)] resolution. To determine an appropriate mapping scale from a known spatial resolution, use the following formula:

Map Scale = Raster resolution (in meters) * 2 * 1,000

Two, in the equation above, is the minimum number of pixels required to detect something. If you have an image of 1-meter resolution, you can detect features at a map scale of 1:2,000 using this formula (1 × 2 × 1,000). This may be helpful if you need to acquire satellite imagery to digitize vector data layers against, or you already have imagery and need to know the scale of map in which it can be used.

When selecting an appropriate spectral resolution, consider the speed of the Internet connection you will use to download the imagery. Higher spatial and spectral resolutions often mean larger file sizes. Consider also the location of the bands in the electromagnetic spectrum, the number of bands, and the number of bits in the image.

As there is more to resolution than simply spatial resolution, there is more to data quality than resolution alone. You must also consider the decisions made by the data collectors that may have compromised the data's resolution and will influence your own decisions. As smaller cell sizes mean larger overall file sizes, slower display, and longer data processing, you may have to sacrifice resolution when assessing the costs and benefits of using a particular dataset. Choosing an appropriate cell size is often a compromise between the conflicting requirements for high spatial resolution versus data storage and processing speeds. Furthermore, cost considerations may force you to consider using public domain satellite imagery, which may be of a lower spectral, temporal, or radiometric resolution than you may prefer for your project.

Recall the discussion in the previous chapter about data quality. With regard to raster data, similar trade-offs exist—particularly among spatial, temporal, spectral, and radiometric resolution—that must be taken into consideration when engineers design a new sensor. These decisions, in turn, affect you the data user. For high spatial resolution, the sensor has to have a small IFOV. However, less energy can be detected with a small IFOV, resulting in a reduced ability to detect fine energy differences and a lower radiometric resolution. To increase the amount of energy detected (and therefore, the radiometric resolution) without reducing spatial resolution, the wavelength range detected for a particular channel or band has to be broadened. However, doing this would reduce the spectral resolution of the sensor. Conversely, coarser spatial resolution allows improved radiometric and/or spectral resolution. In considering the advantages and disadvantages of these three resolutions, they must be balanced against the capabilities of the sensor. In other words, your choice of raster data will inevitably be a compromise as to which resolution you consider most important for a given application. You can't have it all!

Public domain raster data

Having determined the data requirements of your project and identified a need for some public domain raster data, where do you start to look? You learned in chapter 1 that public domain spatial data are available from a number of portals. There have been traditionally fewer portals providing access to raster data than for vector data, largely because of the expense of gathering raster data, in particular, aerial photographs and satellite images. However, as we shall see in this chapter, the volume and diversity of raster data sources are rapidly improving. Overall, there may be fewer raster datasets than vector datasets available, but in terms of the size of the datasets, raster data far exceeds vector data. For example, file sizes may be hundreds of megabytes or even several terabytes for a single hyperspectral satellite image.

You now know there are several key issues to consider when choosing satellite imagery for any GIS project:

- What is the spatial, spectral, radiometric, and temporal resolution of the image?
- Is the resulting imagery panchromatic (an image, black and white or color, usually in the visible part of the spectrum spanning multiple bands), multispectral (many bands), or hyperspectral (up to hundreds of very narrow spectral bands)?
- What type of orbit does the satellite have? Some orbits are geostationary (over a fixed point above the earth's surface), while others are low-earth, polar, or sun-synchronous (polar orbits that are synchronous with the sun so that the surface

is always illuminated by the sun at the same angle no matter where on the earth the image was captured).

- What is the swath of the satellite?
- Is the sensor passive (detecting reflectance of natural radiation from the earth) or active (generating its own signal, bouncing it off the earth's surface and detecting the reflected signal)?
- Is the image already georegistered, and if so, what projection is it in?
- What format is the image in and how much processing is necessary to bring it into a GIS environment?
- How much does the image cost?
- Is the image in the public domain?
- What restrictions are there on the use of the image?

FOR FURTHER READING

While many remote sensing guidelines are available online, the two major remote sensing tutorials are Nicholas Short's tutorial available on NASA's website http://rst.gsfc.nasa.gov/ and from the Natural Resources Canada website, which hosts a series of Earth Observation Tutorials. These guidelines will help you make the right decision as you choose the imagery source that best fits the requirements of your projects.

Key public domain global raster datasets

As with vector data, before you begin working with raster data you should be aware of the types of data that are available so you can search for, select, and download the most appropriate dataset(s) for your project. You may need to download a larger amount of data or additional datasets for your project, as the phenomena you are studying may be connected to other phenomena in the same area. In the following sections, we will discuss the major categories of global and regional spatial data that are considered core to GIS analysis and the sources for those data.

Image data

Image data marked the first widely available public domain dataset for GIS users. In the United States, aerial photographs, initially available on film and paper, were converted to digital format during the 1960s by federal agencies. During the 1970s, digital imagery from Landsat satellites provided the first global coverage. Over the past fifteen years, the amount of imagery available for GIS users has mushroomed—initially from government agencies, but increasingly during the last decade from private companies, as well. This was made possible with the passage of the Land Remote Sensing Policy Act of 1992, which permitted private companies to enter the satellite imaging business.

Landsat

Landsat I, or Land Satellite, launched in 1972 as a result of collaboration between the US Geological Survey (USGS) and NASA, was followed by series of launches culminating with Landsat 7 in 1999. In addition to the source imagery, other raster data were derived from these images, such as elevation and land cover. While other satellites gathered data on weather and oceans, Landsat was the first specifically designed to gather data about the earth's land surface. Three Landsat satellites were still in operation into 2010, providing nearly forty years of continuous coverage for GIS analysts studying change over the earth's surface.

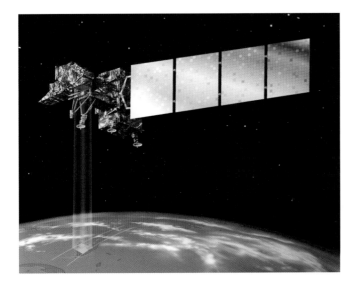

The Landsat satellite, a result of collaboration between the USGS and NASA. Courtesy of NASA Goddard Space Flight Center and US Geological Survey.

Lena River Delta, Russia, imaged by Landsat.
Courtesy of USGS National Center for EROS and NASA
Landsat Project Science Office.

The first three Landsat satellites carried little-used video cameras, and the first five carried the Multispectral Scanner System (MSS), which could sense five bands, including some in infrared, with a spatial resolution of about 80 meters. Later Landsat satellites carried the Thematic Mapper (TM) and Enhanced Thematic Mapper Plus (ETM+) sensors that were able to detect a greater number and different bands than MSS with better geometric quality, increased radiometric resolution, and spatial resolutions down to 15 meters. The TM sensor is a fixed back-and-forth scanner, which senses in eight bands, and is a multispectral scanning radiometer capable of providing high-resolution imaging information of the earth's surface.

With temporal resolutions in the range of sixteen days to eighteen days, Landsat data are available by individual scenes shaped like parallelograms due to the orbit of the satellites and the rotation of the earth. These scenes are referenced in the Worldwide Reference System that divides the earth into paths, corresponding roughly to north-south bands, and rows, which in turn correspond to the center line of latitude of an image scene. Due to changes in the orbits of Landsat satellites, Landsats 1 through 3 use the Worldwide Reference System-1, while Landsats 4 and higher use Worldwide Reference System-2. If you wish to use this imagery, be aware of the different reference systems because they affect how to order and download specific datasets.

A scanning mirror failure in 2003 led to a degradation in ETM+ imagery on Landsat 7 during that year. As a result, approximately 30 percent of the earth is currently not sampled by this satellite. The Landsat Data Continuity Mission continues to pursue additional Landsat satellites, including Landsat 8, but by the time the next satellite is funded and placed into orbit, it may be beyond the estimated life of the currently operating satellites. As a result, the forty-year record of continuous Landsat data may be broken.

Landsat data have always been in the public domain, but were often beyond the reach of many data users due to the price tag of US$600 per scene. Not surprisingly, the announcement in late 2008 that all Landsat data would be free was welcome news for data users. Downloading and using Landsat data subsequently mushroomed. Landsat data are available on three major USGS portals, as well as mirrored on other major data portals, as we shall see in chapter 5. From the USGS Global Visualization Viewer (GloVis) site, users can search for data by clicking an interactive map or by entering the geographic coordinates (latitude and longitude) of an area of interest. After clicking the map or entering coordinates, the GloVis view window will appear on your screen and display nine adjacent Landsat 7 browse images, allowing you to visually locate the best Landsat image for your purposes. A menu at the top of the view window allows you to view data from different sensors, as well as change the browse image resolution and view different map layers. Use the Add and Order buttons, located at the lower left corner of the viewer, to order data; if a Downloadable label appears on the image, the data can be immediately downloaded. Unlike in the past when much Landsat data were unrectified and in formats that were difficult to use in a GIS without remote sensing or conversion software, all data are now orthorectified and provided in GeoTIFF format.

The USGS Earth Explorer site supports the customization of search parameters for Landsat data. Data can be searched by selecting a region on a map, by entering coordinates, or by entering a place name. A specific Landsat dataset, such as Landsat 4–5 TM, Landsat 7 ETM+, Landsat Orthorectified ETM+, and so on, may be selected. The search may be further refined by specifying parameters, such as acceptable cloud-cover percentages and the range of desired acquisition dates. Some Landsat scenes are not archived or distributed by the USGS EDC. These scenes must be ordered directly from the ground stations that acquired the data; pricing and formats vary depending on the data provider. A list of active Landsat ground stations is maintained on the USGS website.

The USGS also offers the Landsat Orthorectified data collection of over 16,000 images for free. This consists of a global set of high quality, relatively cloud-free orthorectified MSS, TM, and ETM+ imagery from Landsats 1–5 and 7. This dataset was selected and generated through NASA's Commercial Remote Sensing Program—a cooperative effort between NASA and the commercial remote sensing community—to provide users with access to global coverage, high-resolution satellite images covering the earth's land masses.

Another enduring archive of Landsat data is NASA's Zulu server. These files are in the Multi-Resolution Seamless Image Database (MrSID) format. This format, developed by a company called LizardTech, works well with large satellite images using lossless compression techniques. (MrSID files can be read directly by ArcGIS for Desktop.) The data on the Zulu server may be downloaded and viewed through the navigation of image tiles, as illustrated in the following example.

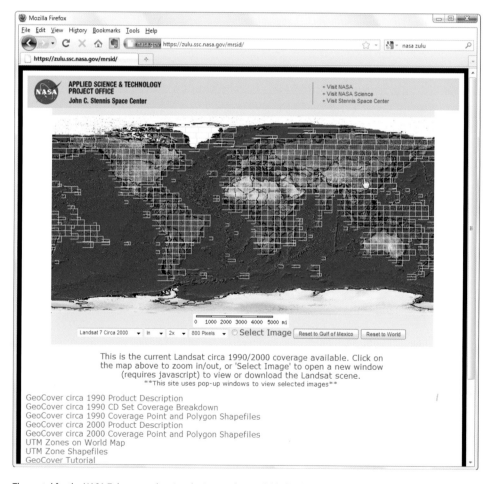

The portal for the NASA Zulu server showing the image tiles available for download. Images provided through NASA's Scientific Data Purchase Project and produced under NASA contract by Earth Satellite Corporation.

Landsat data resides on other sites for specific applications, such as the Global Forest Resources Assessment, a program of the Food and Agriculture Organization of the United Nations. As the name implies, the project is a worldwide assessment of forests and its site includes 400 tiles of Landsat data.

No matter which archive you use for Landsat or other remotely sensed imagery, you should be aware of the data format. If possible, try to choose those file types that require a minimal amount of formatting to bring them into a GIS environment. Even more important, though, is the suitability of different file types for your needs. For example JPEG (Joint Photographic Experts Group) format, a common image compression format, should be avoided for earth surface analysis such as land cover classification because JPEG uses a lossy compression algorithm. The practical implication of this is that pixel values as stored in a JPEG-formatted file are not guaranteed to be identical to the data values recorded by the sensor. Furthermore, while many sites offer satellite image scenes, most are browse images that are only available at a greatly reduced resolution, compared to the original scene at full resolution. Browse images are quick and easy to view through a web browser, but are too low in resolution to support analysis with remote sensing or GIS software. Some data portals offer satellite images, but only in certain bands. Other considerations include the cloud cover of a scene, any postprocessing performed on the scene, and the scene map projection, if any. The ideal satellite image source is one that supports the downloading of all bands recorded by the sensor, at their full resolution, and in a format such as MrSID or TIFF (tagged image file format) that may be accessed directly by ArcGIS for analysis.

FOR FURTHER READING

To learn more about digital image processing, see John Jensen's book, *Introductory Digital Image Processing: A Remote Sensing Perspective* (Prentice Hall 2004), or *Digital Image Processing* (Prentice Hall 2008) by Rafael Gonzalez and Richard Eugene Woods.

SPOT

During the early 1980s, the government of France developed the series of satellites named Système Pour l'Observation de la Terre (SPOT) and launched SPOT-1 in 1986. SPOT was the first satellite system placed in orbit specifically to accommodate the commercial interests not covered by Landsat. To provide a continuous resource of high quality image data, all SPOT satellites provide two types of images, panchromatic and a high-resolution visible. The resolutions for SPOT-4 and SPOT-5 increased to 10 and 2.5 meters respectively, and a mid-infrared band was added. As this combination provides high resolution over large areas, SPOT data are regularly used in resource management and urban planning applications. The sensors can point to areas up to 27 degrees on either side of the satellite path, collect stereo pairs for the generation of elevation data, and revisit an area between one to five days. SPOT data are available from SPOT's corporate website. SPOT's goal is to provide "on-the-spot" service with worldwide availability by leveraging more than thirty direct receiving stations, all handling images acquired by the SPOT satellites.

MODIS

The Moderate Resolution Imaging Spectroradiometer (MODIS) is a NASA research system that collects data from the visible through thermal infrared bands. One of the chief advantages to MODIS data is the fact that thirty-six bands are collected, from 0.4 μm to 14.4 μm. Its temporal resolution is one to two days, and its spatial resolution ranges from 250 meters to 1 kilometer, depending on the band. While only two bands are collected at the 250-meter resolution, in the red and infrared portions of the spectrum, these bands are the most heavily used of all MODIS bands. MODIS data come from two different satellites, Terra and Aqua. The two satellites differ only in the orbit they follow, resulting in different viewing and cloud-cover conditions for a given location. These differences are negligible except for specific applications, such as the Normalized Difference Snow Index used to map snow cover depth and extent.

The raster data collected include albedo (surface reflectance), land cover type, surface temperature, and other products. Many of the raster format environmental data sources that are available today, such as land cover, land use, and vegetation, are largely based on MODIS data. MODIS data may be obtained through the USGS Land Processes Distributed Active Archive Center (LP DAAC).

ASTER

The Advanced Spaceborne Thermal Emission and Reflection Radiometer (ASTER) is an on-demand sensor on the Terra satellite, launched in 1999. ASTER provides imagery in fifteen spectral bands, from visible to long-wave infrared, with a resolution of between 15 to 90 meters. Raster data from ASTER include polar surface and cloud classification, surface emissivity, surface temperatures, surface radiance, and elevation. Users can ask ASTER to acquire specific new data for a small fee (less than US$100) via a request form on the NASA website. Existing data may be obtained from the USGS LP DAAC website. MODIS and ASTER data are also downloadable from NASA's Warehouse Inventory Search Tool (WIST) and at the USGS GloVis website.

AVHRR

The Advanced Very High Resolution Radiometer (AVHRR) is a sensor on board the National Oceanic and Atmospheric Administration's (NOAA) Polar Orbiting Environmental Satellite series. The AVHRR sensor is a broadband, 4- or 5-channel scanning radiometer, which senses in the visible, near-infrared (near-IR), and thermal infrared (or mid-IR) portions of the electromagnetic spectrum. The data collected by the sensor are maintained at the USGS EROS Data Center. The AVHRR instrument provides radiance data to support the investigation of clouds, land-water boundaries, snow and ice extent, ice or snow melt inception, day-and-night cloud distribution, temperatures of radiating surfaces, and sea surface temperature. It may also be used for monitoring vegetation conditions in ecosystems, including forests, tundra, and grasslands, with applications that include agricultural assessment, land cover mapping, image maps, and evaluation of snow cover. The AVHRR imagery can be downloaded from the USGS or from the Global Land Cover Facility website.

VEGETATION

Besides MODIS, another widely used medium-scale resolution sensor is VEGETATION. VEGETATION 1 was placed into orbit in 1998, with VEGETATION 2 following in 2002. These 1-kilometer-resolution sensors were developed by the national space agency of France, the Centre National d'Etudes Spatiales (CNES), which is responsible for shaping and implementing France's space policy in Europe. Both sensors are available on the

SPOT website. CNES data are not in the public domain, although VITO (Flemish Institute for Technological Research Vision on Technology) hosts the entire VEGETATION archive online (for data older than three months) for research purposes. VITO is a research organization, based in Belgium, that provides innovative technological solutions and support to stimulate sustainable development. Scientific users and program partners may obtain VEGETATION products for the cost of the medium (CD or DVD), packaging, and shipping. Orders from research or educational organizations or individuals are processed, subject to the approval of a request, "according to availability of production capacity." Other users must pay for the data, up to €300 for between 1 and 4 million square kilometers. Most of the data are provided in HDF (Hierarchical Data Format), a commonly used raster format.

THE HDF GROUP

The nonprofit HDF Group, housed at the University of Illinois, developed HDF to assist users in the transfer and manipulation of scientific data across diverse operating systems and computer platforms. HDF supports n-Dimensional scientific data arrays, tables, text annotations, several types of raster images and their associated color palettes, and metadata. The HDF software library supports conversion to file types that are more GIS-friendly. Spatial data also are sometimes found in HDF-EOS (Hierarchical Data Format–Earth Observing System) format, which offers the ability to encode location data.

Other NASA data

NASA provides other satellite image data via one of the best resources for the GIS user, the NASA Earth Observations (NEO) website. These data, from a variety of sensors, are categorized into oceans, atmosphere, energy, land, and life, and are available as full resolution georectified images. The themes of these sensors include sea surface temperature, water vapor, net radiation, average land-surface temperature by day or night, solar insolation, permafrost, vegetation index, chlorophyll concentration, and much more. The images are available for a variety of different time periods, as well.

Other government-sponsored satellite imagery

Following the success of Landsat and SPOT, other earth-observing satellite programs proliferated, offering a range of sensor and orbit types. As building, launching, and maintaining a satellite requires large amounts of capital, national governments have traditionally taken the lead in developing remote sensor systems. However, even national governments have enlisted partners from private industry to raise the required funds. For example, the Canadian RADARSAT program, developed during the mid-1990s, was estimated to have cost the Canadian government Can$620 million. The government itself contributed Can$500 million, four participating provincial governments contributed Can$57 million, and the private sector contributed Can$63 million, although none of this included the launch of the satellite. Most governments find it advantageous to use a private company to market and distribute the satellite imagery; in the case of RADARSAT, it is RADARSAT International, through MDA (MacDonald, Dettwiler and Associates Ltd.). In addition, as organizations need to recover the high cost of these satellite programs, some of the cost is passed on to the data user. For example, one 3-meter RADARSAT scene cost Can$5,400 and one SPOTView ortho scene cost €2,500 in 2009. Not surprisingly, most of these satellite image datasets are not in the public domain.

The RADARSAT program saw the launch of its first satellite, RADARSAT-1, in 1995, followed by RADARSAT-2 in 2007. As the name implies, these sensors image the earth in the microwave radar region of the electromagnetic spectrum at a single microwave frequency of 5.3 gigahertz in the C band with a wavelength of 5.6 centimeters. RADARSAT uses a Synthetic Aperture Radar (SAR) sensor to transmit microwave energy towards the earth's surface and, unlike optical satellites that sense reflected sunlight, SAR systems transmit microwave energy towards the surface and record the reflections. This means RADARSAT-1 can image the earth, day or night, in any atmospheric condition, including cloud cover, rain, snow, dust, or haze.

The European Space Agency (ESA) launched ERS-1 and ERS-2 with radar systems in the 1990s, and then in 2002 launched Envisat, with both spectral imaging and radar and a wide-angle capability. Sample Envisat data are available to download.

Russia's Resurs series offers Landsat-like imagery with three visible and near-infrared bands, two thermal bands, and a spatial resolution of 160 meters for visible/near-infrared, and 600 meters for thermal (mid-infrared). The data are available on the Russian Federal Space Agency Research Center for Earth Operative Monitoring website. Resurs has operated for more than forty years, much of that time exclusively for military purposes. Some data are retrieved from physical film media, ejected from the satellite attached to a parachute. Since 1992, the National Space Agency of the Ukraine has worked in cooperation with the Russian Federation, and its OKEAN series includes multispectral scanners, thermal sensors, and radar.

The Indian Space Research Organization operates several earth-resources satellites that gather data in the visible and near-IR bands. Its first satellite, IRS-1A, was launched in March 1988. IRS-1D, launched in 1997, captures data in the blue-green, green, red, and near-IR bands at a 23-meter spatial resolution. This produces 5.8-meter panchromatic images and 188-meter resolution wide-field (large area) WiFS multispectral images. In 2003, ResourceSat-1 began producing multispectral images at 56-, 23-, and 5-meter resolutions. Japan's Aerospace Exploration Agency began the JERS (Japanese Earth Resources Satellite) program in 1990, which includes both optical and radar sensors.

NASA's ALOS (Advanced Land Observation Satellite) was successfully launched on January 23, 2006. The ALOS has three remote sensing instruments: the Panchromatic Remote-sensing Instrument for Stereo Mapping (PRISM) for digital elevation mapping, the Advanced Visible and Near-Infrared Radiometer type 2 (AVNIR-2) for precise land coverage observation, and the Phased Array type L-band Synthetic Aperture Radar (PALSAR) for day-and-night and all-weather land observation. Perspective scenes have been produced by combining digitized land elevation data with a corresponding AVNIR image.

The Peoples Republic of China joined forces with the Brazilian government to develop a series of earth-observing satellites launched by Long March rockets from China. The satellites are referred to as the CBERS (China-Brazil Earth Resource Satellites), or in China as the Zujuan series. CBERS-1 launched in 1999 and includes WFI (300-kilometer swath; 260-meter resolution; 4 bands), IRMSS (20-kilometer swath; 80-meter resolution; 4 bands, including thermal), and CCD (20-meter resolution; 4 bands). CBERS-2 (YZ-2) was launched in 2000 and its 3-meter resolution sensor is being used primarily for military reconnaissance rather than for earth imaging. Most images cover Brazil and China and are not generally available to other nations. Requests for access to the data may be made via the Brazilian Ministry of Science and Technology.

The United Kingdom has developed a series of small (440 kilograms) but versatile satellites, which, because of their size, are referred to as microsatellites. These include a single sensor that can image in the green, red, and near-IR wavelength band intervals. The microsatellites are used in a configuration that allows any location on earth to be visited once during a 24-hour period. Such an arrangement is especially useful to monitor disasters, covering an area 600 km × 600 km at a resolution of 34 kilometers. The first satellite (AlSat-1) was launched in 2002, and collectively these satellites form the Disaster Monitoring Constellation. The imagery is distributed by DMC International Imaging Ltd.

Primary private sources of satellite imagery

The two main private sources of global satellite imagery based in the United States are GeoEye and DigitalGlobe. GeoEye, which began as OrbImage and acquired Space Imaging in 2006, launched the first commercial sub-meter satellite, IKONOS, in 1999. Its other satellites include OrbView-2 and 3, and GeoEye-1.

DigitalGlobe operates QuickBird (launched in 2002, the world's first 60-centimeter-spatial-resolution satellite), WorldView-1 (launched in 2007), and WorldView-2, launched in 2009. WorldView-2 houses a 46-centimeter panchromatic and an 8-band 184-centimeter multispectral sensor. Increasingly, these image types are used for generating derivative spatial data. For example, twelve different land cover types can be automatically generated from the WorldView-2 satellite imagery.

By 2010, DigitalGlobe's content library contained more than one billion square kilometers of earth imagery, 33 percent of which was reported to be less than one year old. This company alone adds approximately 1.5 million square kilometers of imagery every day—roughly the size of Saudi Arabia, or three times the earth's landmass annually (GIM International 2010). Some of DigitalGlobe's imagery is offered through eMap International.

The competition between these companies has resulted in an amazing improvement in the quantity and quality of imagery available for a variety of applications, ranging from GIS and remote sensing, to everyday web mapping tools such as Yahoo! Maps.

ArcGIS Online

Using ArcGIS Online, you can find, use, and share geospatial data, tools, and applications. It also serves as a source of imagery for GIS applications. The online option offers imagery for free, as well as via a subscription service. A DataDoors service allows users to find, purchase, and download imagery straight into ArcGIS. The Data Appliance for ArcGIS allows users to purchase a hardware device containing pre-loaded content to use on a local network.

The World Imagery dataset on ArcGIS Online contains both global satellite imagery and high-resolution aerial imagery for the United States, and hundreds of cities around the world. It includes NASA Blue Marble's 500-meter resolution imagery at small scales (at coarser than 1:1,000,000), 15-meter eSAT imagery at medium-to-large scales (to 1:70,000) for the world from i-cubed (the Fort Collins, Colorado, imagery company), and USGS 15-meter Landsat imagery for Antarctica. It also includes 1-meter i-cubed imagery for the continental United States and GeoEye IKONOS 1-meter resolution imagery for Hawaii, parts of Alaska, and several hundred metropolitan areas around the world. The i-cubed data are a seamless color mosaic of various government imagery sources, including USGS imagery for metropolitan areas, the best available USDA National Agriculture Imagery Program (NAIP) imagery, and enhanced versions of USGS digital orthophoto quarter quadrangle (DOQQ) imagery for other areas. The data also include imagery assembled by Esri from various sources through an online content sharing program.

This imagery is licensed under a Creative Commons Attribution-NonCommercial-ShareAlike 3.0 United States License. This allows anyone using the data to share (copy, distribute, and transmit the work) and remix (adapt the work) for free, as long as the user attributes the work in the manner specified by the author or licensor but not in any way that suggests that the author or licensor endorses the data user or the use of the work. The data must also be properly cited, including metadata, so that others may locate the original work. The data may not be used for commercial purposes, and if the data are altered or built upon, the data user may only distribute the resulting work under the same or similar license to this creative commons agreement. These conditions can be waived if the data user obtains permission from the copyright holder.

Another set of imagery on ArcGIS Online is the United States Prime Imagery map service. This provides i-cubed Nationwide Prime high-resolution (1 meter or better) aerial imagery for the contiguous United States. A private imagery firm, AEX Aerials, provides i-cubed Nationwide Prime, a seamless, color mosaic of various commercial and government imagery sources, including 0.3-meter to 0.6-meter resolution imagery for metropolitan areas. For other areas, the map service is composed of the best available USDA NAIP imagery and enhanced versions of USGS DOQQ imagery. Unlike the world imagery data, these data are copyrighted by the license agreement that accompanies the use of ArcGIS software, largely because much of the prime imagery comes from private third-party sources. The spatial resolution of imagery on ArcGIS Online improves by the day. We will return to online uses of ArcGIS in chapter 9 when we discuss spatial data in the cloud.

Other raster image data

Despite the plethora of available satellite and aerial imagery over the past decade and the anticipated increased use of such imagery over the next decade, another category of imagery has the potential to far exceed traditional overhead or oblique public and private imagery. Over the past few years, attaching geographic coordinates or *geotagging* other data not previously considered a GIS resource has introduced an enormous amount of information to what is now referred to as the "GeoWeb" and "Citizens as Sensors" (Goodchild 2007, see reference and additional discussion in chapter 8 on crowdsourcing).

A high proportion of this additional data consists of photographs (and increasingly video) taken on the ground and often available in real-time from static or mobile cameras. These photographs and videos are finding their way onto public websites such as Panoramio, Flickr, and PicasaWeb. Although they are becoming increasingly popular as a source of imagery data, not all this information is available in the public domain and not all is of the necessary quality to use in conjunction with other geospatial information for geographic information analysis.

Some data have been collected by private vendors, such as Google, whose StreetView product provides street-level static images linked to specific locations along thousands of kilometers of roadways around the world. This is being extended to include pedestrian walkways, trails, and bikeways as well. Although publicly available, anyone wishing to make use of this information should pay careful attention to the distribution rights and use restrictions governing Google imagery.

The future of raster image data

As with other geo-technologies, the world of remotely sensed imagery is rapidly changing. For example, in 2010 and 2011, SPOT launched two sub-meter Pleiades satellites with a multispectral spatial resolution of 2 meters and a panchromatic spatial resolution of less than 1 meter. GeoEye is launching GeoEye-2 with a spatial resolution of 25 centimeters. As the demand increases for imagery across all fields, private and government providers will continue to offer increased levels of coverage and detail, resulting in more data for GIS users. However, many of these datasets are likely to involve some cost to the end-user and some restrictions as to how the data may be used. In chapter 9, we will discuss the impact of new raster imagery and changing technologies on solving problems with public domain data in the future.

Digital elevation data

DEMs (digital elevation models) are a form of raster data in which each cell contains a value that is most representative of the elevation in that cell. The size of the cells can range from less than 1 meter to more than 1 kilometer, depending on the method used to collect the data. The earliest DEMs were derived from topographic maps by scanning the contour lines originally stereocompiled from pairs of aerial photographs.

DEMs from national mapping agencies

DEMs can be produced from large-scale data at less than 1 meter to 5 meters for small areas, from 5- to 50-foot contours from USGS topographic maps, or from 25-meter contours from Ordnance Survey maps. DEMs find application wherever data about elevation, slope, direction of slope, viewsheds, or other information about the earth's terrain are required. They are widely used in mitigating the results of landslides, siting ski areas, planning forest management and logging activities, assessing military landing sites, studying soil erosion, and in many other applications.

Although a valuable resource for terrain analysis, DEMs are not without their limitations. Topographic maps use a wide variety of contour intervals, depending on the amount of surface relief in a particular area, and different methods have been used to create topographic maps over the decades. As a result, contour lines, and the DEMs derived from them, may not match along their edges. Also, DEMs were not originally created with an appreciation of water features; consequently, lakes may not appear as a smooth surface, and rivers may appear not to flow through the lowest points in a valley. In the United States in 2000, these anomalies led to the creation of the National Elevation Dataset (NED). This dataset standardized the horizontal and vertical datums, matched edges, integrated hydrography and human-made features, and incorporated other enhancements. NED data represent the highest resolution national coverage of DEMs in the United States and were originally available at a resolution of 1 arc-second (30 meters) for the entire country, except for most of Alaska, which is available at 60 meters. Since 2000, more of the country has been mapped as $^1/_3$ arc-second (10 meters) resolution data and small areas are mapped as $^1/_9$ arc-second (3 meters) data.

As USGS DEMs have been available in the public domain for the past twenty years, they were one of the first government public domain datasets to be mirrored (or copied) and made available on a plethora of federal government portals, state GIS portals, and private vendor sites. The federal government portals include the USGS Seamless Data Distribution System and the USDA's Natural Resources Conservation Service's Geospatial Data Gateway. These portals both provide access to the data via web mapping interfaces. Examples of portals implementing a more traditional interface—a list of map names that correspond to latitude-longitude blocks of data—are the USGS Geographic Data Download and Earth Explorer sites. These more traditional resources may seem outdated, but they are a useful resource when bandwidth is limited. They can often support faster download speeds when you know exactly which latitude-longitude block is required.

States offering DEM data include Nebraska and Tennessee. Some of the private vendors, such as MapMart, charge a fee for the data, while others, such as the Geo-Community's GIS Data Depot, provide free downloads. Other sites, such as the Digital Atlas of Oklahoma hosted by the USGS, provide access to products derived from DEMs, the most common being hillshades, slope maps, and aspect (direction of slope) maps. Derivative products are beginning to disappear from online sites, perhaps because they are so easy to generate with GIS software, such as the Spatial Analyst extension in ArcMap.

Digital elevation models were first offered in a text-based DEM format that required the GIS user to run a DEM-to-grid conversion tool in ArcGIS. While some sites still offer this format, more sites offer the DEMs in an ASCII Grid format, which requires a different tool (ASCII-to-grid) to convert the data to an ArcGrid. Other sites offer DEMs in a *smart image* format, such as GeoTIFF, while others offer DEM data in already-processed rasters. Some DEMs are stored with integer elevation values, some with decimal elevation values, and some, including the Montana Natural Resources Information System, provide them in both formats. Pay attention to the vertical datum used for the elevations in the DEMs and whether the values are in feet or meters.

GTOPO30 and ETOPO1

The intensive labor required to create DEMs from aerial photographs or topographic maps for local areas, regions, or countries motivated the GIS user community to seek a global DEM. The first such dataset, still in use today, was GTOPO30. This was completed in 1996 as a collaborative effort of NASA, the USGS, the United Nations Environment Programme (UNEP), United States Agency for International Development (USAID), Instituto Nacional de Estadística y Geografía (INEGI) of Mexico, the Geographical Survey Institute of Japan, Manaaki Whenua Landcare Research of New Zealand, and the Scientific Committee on Antarctic Research (SCAR). The *30* in the name of this dataset comes from its horizontal grid spacing of 30 arc seconds, approximately 1 kilometer. The dataset was generated from sources that included digital terrain elevation data, the Digital Chart of the World, USGS DEMs, US Army Map Service maps, the International Map of the World, and sources in Peru, Mexico, Japan, New Zealand, and Antarctica. GTOPO data are available from the USGS Earth Resources Observation and Science Center.

ETOPO1 is a 1 arc-minute global relief model of the earth's surface, integrating land topography and ocean bathymetry. It was built from different global and regional datasets and is available in a surface depicting the top of the ice sheets in Antarctica and Greenland, and a surface depicting the bedrock elevation. It was created by the NOAA National Geophysical Data Center to replace the ETOPO5 5-minute gridded dataset and is available on NOAA's website in several formats, including NetCDF (Network Common Data Form), binary float, xyz, and the easiest format for GIS users—a georeferenced TIFF file.

SRTM

By 2000, increasing requirements for higher resolution data led to funding for the direct generation from space of a DEM for the entire planet. The Shuttle Radar Topography Mission created the first global DEM data at 30-meter resolution. This was accomplished from a radar signal shot from the end of a long 60-meter mast carried by the Space Shuttle *Endeavour*, which was received by a sensor in the shuttle's bay through a technique known as interferometric synthetic aperture radar (IFSAR). Only very small areas in high mountains and deep gorges (less than 0.2 percent of the planet) were not able to be collected using this technique. The data are distributed at 1-arc-second (30 meter) resolution for the United States and its territories and, for the rest of the planet, at 3-arc-second (90 meter) resolution. SRTM data are delivered through the USGS Seamless Data Distribution System and USGS Earth Explorer. A plain text website for the data is also available. The remote sensing archives, Global Land Cover Facility and the CGIAR Consortium for Spatial Information, also host SRTM data.

REDUCED RESOLUTION OUTSIDE UNITED STATES

Although SRTM data were collected at 30-meter resolution for the entire planet, they are distributed at a reduced resolution of 90 meters for everywhere outside the United States. Do you think the reason for distribution at the reduced 90-meter resolution was to protect sales of elevation data created by national mapping agencies, for national security, or another reason?

SRTM data files have names such as N34W119, with each data file covering a 1 × 1 latitude-longitude degree block of the earth. The first seven characters of the file name indicate the latitude and longitude of the southwest corner of the block, with N, S, E, and W referring to north, south, east, and west, respectively. The file N34W119.hgt covers latitudes 34 to 35 North and longitudes 118 to 119 West. The filename extension hgt simply represents height, meaning elevation; it is not a format type. These files are a raw format. They are uncompressed, 16-bit signed integers, with elevation measured in meters above sea level, in a geographic (latitude and longitude) projection, with data voids indicated by the value -32768. United States 1-arc-second files contain 3,601 columns and 3,601 rows of data, with a total file size of 25,934,402 bytes (3,601 × 3,601 × 2). The international 3-arc-second files contain 1,201 columns and 1,201 rows of data, with a total file size of 2,884,802 bytes (1,201 × 1,201 × 2).

More recently, NASA released version 2 of the SRTM data, also referred to as the "finished" version. Version 2 is the result of a substantial editing effort by the US National Geospatial-Intelligence Agency (NGA), which has produced well-defined water bodies and coastlines, and removed spikes and wells (errors in individual pixels). Some areas of missing data, or *voids*, are still present, however. The version 2 directory also contains the vector coastline mask derived by NGA during the editing process, called the SRTM Water Body Data (SWBD) available in Esri shapefile format. The data, version 1 and version 2 and the accompanying metadata, may be obtained through the USGS SRTM website.

An updated version of the GTOPO30 dataset has also been released, with SRTM data used in place of the original data when possible. This dataset is also available on the USGS DDS SRTM server, located in the SRTM30 directory. A summary of the main types of global elevation data available from the USGS is available on the USGS EROS Center website.

ASTER DEMs

During 2009, a collaborative effort between Japan's Ministry of Economy, Trade, and Industry and NASA led to the development of a global digital elevation model based on ASTER data called ASTER GDEM. The ability to produce DEM data resulted from the ASTER satellite containing a stereoscopic imaging capability in the near-IR band. The elevation data are extracted from the stereo images by auto-stereo correlation between the nadir view (directly below the satellite) and the along-track oblique view from the ASTER sensors. The data are partitioned into 1 × 1 degree tiles and are available to download in small blocks, about 8 megabytes to 12 megabytes in size (compressed in a zip file). Uncompressed, the DEM is about 24 megabytes. Upon completing the registration and request forms, the download site, an FTP URL, is e-mailed to the user. Data may be obtained either from the ASTER GDS IMS website or the WIST portal hosted by NASA.

The availability of ASTER DEMs elevation data for the planet, at 30-meter spatial resolution, is welcome news for the data user community. However, bearing in mind a previous warning to understand the data before you use them, use ASTER DEMs with caution because they are not generated in the same way as other digital elevation data. ASTER is a relative DEM in that the elevation data have been constructed entirely using the stereo images, with no corrections from ground control points. The positional accuracy and the feature integrity are determined by the sensor parameter settings and the shape of the earth as measured by the geoid. An accuracy assessment was performed on the ASTER DEMs by comparing it with other DEM and ground control points, and it suggested an overall horizontal and vertical accuracy of approximately 20 meters. Topographic artifacts and inaccuracies in feature geometry also exist. You may be able to compensate for these imperfections in the ASTER data source by inputting additional elevation information from ground control points using a GIS or remote sensing software package and then reconstructing the DEM. In addition, the elevations are posted at 30-meter intervals, but the details of topographic features that can be resolved appear to be approximately 100 meters. Still, given the spatial resolution of this DEM, its availability (for free) in the public domain, and its global coverage, it remains a worthy choice for data analysts.

Lidar DEMs

Light detection and ranging (lidar) is another remote sensing technique employed to collect topographic data. This technology has been increasingly used since the late 1990s by such organizations as NOAA and NASA to document topographic changes along shorelines, and by the US Forest Service to detect changes from wildfire burns. Lidar data are collected by lasers mounted on aircraft, capable of recording elevation measurements at a rate of 2,000 to 5,000 pulses per second or more. One of the main advantages of lidar data is vertical precision; in the late 1990s, it was already around 15 centimeters (6 inches) and has gradually improved since then. Lidar sensors pulse a narrow, high-frequency laser beam toward the earth through an opening in a low-flying (600 meters or so) aircraft. Another advantage is lidar can detect bare earth, vegetation, and the built environment. Although lidar data cannot be collected during times of rain, fog, or high winds because the water vapor and air currents could cause the laser beams to scatter and give a false reading, the data can be collected at night or through clouds. A moving mirror produces a conical sampling pattern beneath the aircraft over a certain swath, sometimes 30 degrees wide. This allows the collection of DEM data over a strip that is typically a few hundred meters wide. Reflected laser light from the ground is directed onto a small telescope. The lidar sensor records the time difference between the transmission of the laser beam and the return of the reflected laser signal to the aircraft, using the difference in time to calculate the elevations.

However, lidar equipment collects only elevations; it does not record the positions of the data points where the elevation data are captured. To spatially enable the data, these data points must be recorded, so a high-precision GPS is mounted on the aircraft fuselage. As the lidar sensor collects data points, the location of the data is simultaneously recorded by the GPS sensor. After the flight, the data are downloaded and processed, resulting in accurate, geographically registered x, y, and z positions for every data point. These "x,y,z" data points support the generation of a very accurate DEM of the ground surface, which may be used to detect small changes and features, such as the uplift from Mount St. Helens and the location of the fault that runs underneath Seattle, Washington.

Due to the high cost of flying and generating lidar data, very little of it was initially available for free and very little was in the public domain. Now, however, some significant regions of lidar data are available. One portal providing access to lidar data is OpenTopography maintained by the San Diego Supercomputer Center. Data for selected regions of the United States, such as Southern California, are available in DEM, an ASCII-text point cloud data format, and KMZ files that may be brought into ArcGIS Explorer. For most GIS applications, filtered ("bare earth") and unfiltered 0.5-meter resolution DEMs in Arc Binary grid format are also available for download as 1-square kilometer tiles. The site also provides access to web-based tools for processing point cloud data into custom DEMs.

As a result of the expense, limited collection range, and lack of any national or international lidar programs, various consortia have emerged to gather and provide lidar data access to stakeholders and data users alike. For example, the Oregon Lidar Consortium grew from a local effort in Portland to provide lidar data for the state. Among the many benefits of the consortium approach are the following:

- As large swaths of data can be collected seamlessly, the costs per unit area to collect the data are greatly reduced.
- Expert quality assurance and quality control are uniformly applied to the data.
- Statewide standardization of data is assured.
- Small jurisdictions benefit from the acquisition of lidar data at a greatly reduced cost compared to the cost if jurisdictions acted independently.

The Puget Sound Lidar Consortium based in Seattle follows an approach similar to Oregon's. The lidar depositories from the USGS, NOAA, and other federal agencies are also listed on the Lidar Links for Mappers website. Other data repositories, such as the Inside Idaho data portal, are run by GIS clearinghouses in individual states.

Digital raster graphics

One of the simplest, but most frequently used, of all raster datasets are scanned, georegistered topographic maps, often referred to as digital raster graphics (DRGs). The latest edition of all 55,000 USGS topographic maps were originally scanned in the late 1990s at 250 dpi. Public domain DRGs are available on most state data portals, as free data from private vendors, such as the GIS Data Depot, and even grassroots efforts, such as the Libre Map Project.

However, even for something as basic as a GIS-ready version of a topographic map, there are differences in available type and format along with some licensing restrictions. Some sites offer DRGs with the original map collars (or margins), while others provide the data in a variety of projections and with the collars removed. As with other data, the more flexibility with the data you require, the more likely the site will offer the services for a fee. Topographic maps are frequently provided by national mapping agencies around the world but they are rarely in the public domain and usually only available for a fee. Furthermore, 250 dpi is generally suitable as a backdrop image but is unsuitable for publication or for very small areas requiring high resolution.

The popularity and public domain nature of these backdrop topographic images for field data and a host of recreational applications, such as imagery on handheld GPS devices, prompted private companies to rescan these maps at a higher resolution and, in some cases, add shaded relief to the imagery. National Geographic Maps, for example, first offered enhanced DRGs as part of its TOPO! product. Unfortunately, these enhanced DRGs were copyrighted due to the value added and were kept in National Geographic's proprietary format, rendering them unusable in a GIS environment. The USGS now provides its topographic maps as a product called US Topo as GeoPDFs via the National Map portal. GeoPDF is a georegistered Adobe PDF format, and while this format supports measurement and some other basic functions, it cannot be analyzed by GIS software in the same manner as raster DRGs.

Sometimes there may be variations in the original base topographic maps that you should be aware of. For years, the USGS and US Forest Service each produced similar topographic maps for areas covering Forest Service lands, encompassing approximately 20 percent of the United States. The Forest Service maps, called single edition quads, contain more detailed information on road and other features in the forests; only digital versions were produced after 2006. On the USGS portal, georeferenced PDFs are available via the map locator and downloader tool. Occasionally, especially in national forests, the PDF has been superseded by a more recent version from the Forest Service that is not available from this website.

On the Forest Service's ArcIMS portal, the Geospatial Service and Technology Center, a somewhat different map is portrayed. Again, no green vegetation layer is present, while private land is shown as areas with black outline. The date, 2006, is given in the map key grid to the right of the bar scales. There is no 100,000-meter × 100,000-meter Military Grid Reference System zone indicated on these single edition quads. On the other hand, the section-township-range public land survey system appears on the single edition quads, but not on the forest lands shown on USGS topographic maps. Universal transverse Mercator (UTM) grid ticks appear in the margin of the single edition quads,

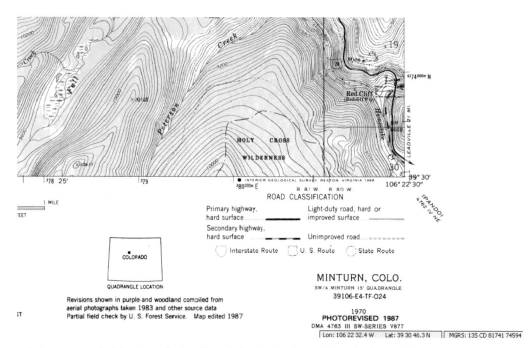

This map shows a DRG from the online USGS Store, indicating what is available as a hard copy USGS topographic map. It was last updated in 1987, before the Forest Service started using a gray tint to denote private land within the forest. Courtesy of USGS.

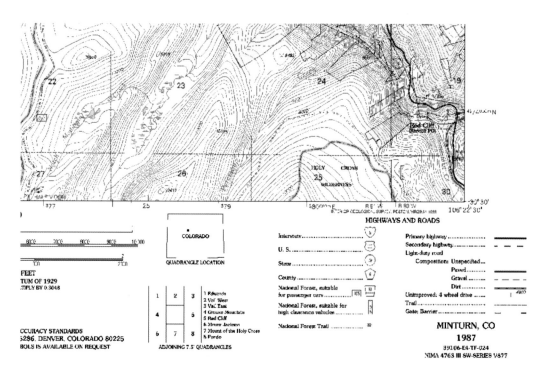

This DRG from the Forest Service site is a lower resolution scan. Although it contains a 1987 date, private land is shown with gray tint, and there is no green vegetation layer. Courtesy of USGS and US Forest Service.

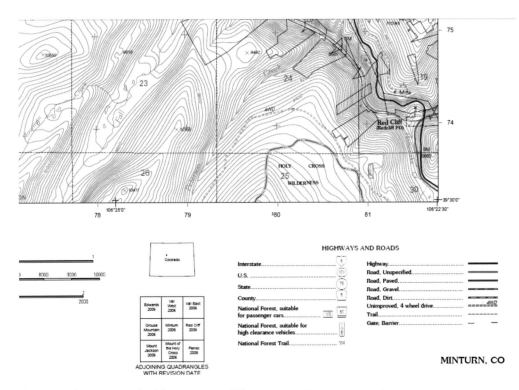

This topographic map was updated from new vector GIS layers. Courtesy of USGS and US Forest Service.

but unlike USGS topographic maps, there is never a full UTM grid displayed across the map.

Increasingly, topographic maps are offered as a layer in online mapping services. For example, during 2008, Esri began offering enhanced DRG images from National Geographic as part of the standard base layers in its online series of basemaps. However, as with the topographic maps offered as individual tiles, all prospective users of the data should take time to understand the content of these topographic maps, who created them, and when they were created. The metadata may be more difficult to obtain for the specific topographic maps in a map service layers as compared to traditional map files, but it is essential to have access to metadata, especially as many vector (such as hydrography) and raster data layers (such as DEMs) are based on content derived from these original topographic maps.

Topographic maps have been produced by national mapping agencies, such as Ordnance Survey and the USGS, since the 1800s. Some of historical editions of the data have been maintained as well, providing useful reference sources for historical studies. For the northeastern United States, for example, historical topographic maps have been available for several years from Maptech, later hosted by MyTopo. Topographic maps and historic aerial photographs are also available as georeferenced JPEG files on the Terraserver USA archive, later to become Microsoft Research Maps (MSR Maps). This portal was established following an agreement between the USGS and Microsoft in 1998. In 2009, the USGS announced it had launched a project to scan and georectify the thousands of historical topographic maps that it has captured at several different scales and make them available to the public. The tens of thousands of maps in the collection may not be available for several years. In the United Kingdom, the Landmark Information Group and Ordnance Survey provide historical maps dating back to the mid 1800s via the Old-Maps website.

Digital orthophoto quadrangles

The collection and use of aerial photography began in earnest during the 1930s, and despite the widespread availability of GIS-ready satellite imagery today, aerial photography continues to be widely used. It is also more likely be available in the public domain than satellite imagery. Raw aerial imagery contains distortions that must be removed before the imagery is used in mapping or GIS-based projects. Radial distortion (which increases as you move away from the center of the photograph), along with the position of the airplane, the relief of the terrain, and the camera lens itself, combine to introduce distortions which render an aerial photograph inaccurate for use in mapping. These distortions are removed to create an *orthophotoquad*, an accurate base upon which road, contour lines, rivers, and other features are compiled to create topographic maps. With the advent of digital cartography and GIS, a digital version of this product, known as a digital orthophoto quadrangle (DOQ), became a standard base layer used in conjunction with, or without, a topographic map.

Aerial photographs are still used to create DOQs when the cost of satellite imagery is prohibitively expensive. Due to the expense of planning and executing aerial photography, such projects are frequently undertaken as a partnership among local, regional, state, and national government agencies, and on occasion, with private industry and nonprofit organizations. An example of this is the NAIP, which usually includes the USDA, the USGS, and state agencies, and gathers data by flying over states (or portions of states), often at 1 meter resolution. Although NAIP imagery can be purchased from the USDA, the imagery is often provided for free on state portals, such as the Texas Natural Resources Information System (TNRIS) and the Virginia Information Technologies Agency (VITA) websites.

A variety of sources exist for historical aerial photographs, but they may not be georegistered or even be available in digital format. Local historical societies and museums are often the best source. Images from the National Aerial Photography Program of the USGS, which was active from the 1980s to around 2000, and the National High Altitude Photography Program, which operated during the 1980s, are available in digital form. Older photographs used to create the original topographic maps may also be purchased, scanned, and georegistered. Even some declassified military satellite images, for example some from the Corona program of the 1950s and 1960s, are available from the USGS.

WHAT'S OLD BECOMES NEW

At the end of 2009, the aerial photo company Bluesky uncovered a series of 1949 original aerial survey photographs covering all of Greater London. The original survey was taken by Hunting Aero Surveys in June and July of 1949 and comprises two sets of photographs: a low-scale survey of all Greater London and coverage of the old administrative County of London at a higher scale. The original contact prints (prints made directly from the film negatives) will be scanned for posterity, georeferenced, and uploaded to the OldAerialPhotos website run by Bluesky International Limited. This site contains over one million photographs, which can be searched using a street address, postcode, or grid reference.

As the market for historical map data grows, more aerial photographs, ground photographs, and paper maps locked away in museum cabinets, historical societies, and government warehouses, will be made available. Whether those data would be available for free depends on the organization that collected the data, who currently maintains the data, and whether they seek to make a profit on those materials.

Satellite imagery versus aerial photography

Given that is it is relatively easy to generate satellite imagery today, why is aerial photography still used? During the 1980s and 1990s, most satellite imagery was of low spatial resolution and comparatively expensive, so aerial photography missions were routinely flown to maintain the supply of higher quality imagery. State and federal agencies in the United States often collaborated on the NAIP to pool resources and generate 1- or 2-meter resolution aerial photography for states or specific regions. By 2010, costs for satellite imagery had dropped substantially, the temporal resolution had improved to weekly or even daily, and there were gradual improvements in spatial resolution. Despite the changes in satellite imagery, aerial photography remains a popular choice.

Some local governments own their own aircraft and aerial cameras or have established a cost effective partnership with a local or regional aerial survey company. Airplanes also have certain flexibility that satellites do not, such as waiting for optimal sky conditions. Governments also have the ability to send multiple airplanes to multiple locations at once. Aerial photography can still acquire higher spatial resolutions than satellite imagery (such as Vexcel's 6.50-centimeter imagery vs. GeoEye-1's 0.41-meter and WorldView-1's 0.50-meter panchromatic imagery). Thurston (2010) pointed out that "we are probably nearing the day when aerial cameras will be combined with lidar into the same device to provide both kinds of data at the same time, something satellite imagery is likely not about to provide for some time to come."

However, the gap is narrowing, particularly with advances in image extraction moving photogrammetric processes away from hardware; into digital workflows, cloud computing, 3D modeling, and virtual reality; and toward services that help users interpret the imagery. Aerial photography for large areas continues to be expensive and requires partnerships among many different agencies to generate enough financial commitment for complete coverage of a region—which is even more important for a long-term program that includes updates. For example, the partners required to support the Imagery for the Nation program in the United States include professional GIS societies such as American Congress on Surveying and Mapping (ACSM), American Society for Photogrammetry and Remote Sensing (ASPRS), and University Consortium for Geographic Information Science (UCGIS), and government organizations such as statewide GIS councils, the National Association of Counties, and the Western Governors Association.

Satellite imagery is becoming an increasingly viable alternative to aerial photography, particularly with the advent of initiatives to cover often-requested areas with imagery as a subscription service. For example, DigitalGlobe and Microsoft formed a partnership in 2009 on a Clear30 program; that partnership uses Microsoft's Vexcel UltraCam digital aerial camera to collect 30-centimeter aerial imagery (in 1 degree × 1 degree cells) of the entire contiguous United States and Western Europe and expects to complete initial collection by 2012. At the time of its launch, the plan was that organizations could sign up for the program that provides them not only with imagery but also with updates, as certain urban areas will be reimaged every year or two.

Land cover data

Land cover spatial data are often a mixture of what is on the land—land cover—and what the land is used for—land use. The data are most typically collected and served in a raster format at a regional, national, and international level, but are sometimes offered as vector area data at the local level. At the regional to international level, land cover data are frequently derived through the interpretation of aerial photography or satellite imagery by supervised, or unsupervised, classification methods using remote sensing software.

The Global Land Cover by National Mapping Organizations (GLCNMO) is one example of a global land cover dataset. These data are available via the International Steering Committee for Global Mapping website as a 1-kilometer-resolution grid (30 arc seconds) containing twenty land cover items. The data were created by using MODIS data observed in 2003 from the Terra satellite. The classification is based on a system called LCCS, developed by the Food and Agriculture Organization of the United Nations (FAO) so that other land cover products based on the same classification system could be compared. Another dataset resulting from the same collaborative effort is the Global Percent Tree Cover. This represents the density of trees on the ground; the ratio of the area covered with branches and leaves or trees (tree canopy) to the ground surface as seen from above. The tree cover dataset was also based on the 1-square-kilometer-resolution satellite images from the MODIS sensor on the Terra satellite. In terms of deciduous trees, which drop all their leaves during the period of low temperature or dryness, the ratio of the most flourishing period of a year (Maximum Percent Tree Cover) is referenced as the Percent Tree Cover. POSTEL in France also makes a global land cover product from 300-meter resolution satellite imagery. In New Zealand,

land use is divided into sixteen regional datasets and is distributed as a file geodatabase by the New Zealand Ministry for the Environment.

In the United States, land cover data collected by the USGS were based on a method pioneered by James Anderson and others that classifies land cover and land use into three hierarchies representing over 100 land use types (Anderson et al. 1976). These types include open water, perennial ice and snow, developed open space, developed low intensity, developed medium intensity, developed high intensity, barren land, unconsolidated shore, deciduous forest, evergreen forest, mixed forest, dwarf scrub, shrub/scrub, grassland, herbaceous, lichens, moss, pasture, cultivated crops, woody wetlands, palustrine wetlands, scrub/shrub wetlands, and estuarine beds. The USGS data, called the National Land Cover Data (NLCD), were collected by examining Landsat data through an accurate but laborious process that requires thousands of hours to manually interpret satellite imagery. The results are a detailed assessment of the earth's land use and land cover. NLCD data were collected in 1992 and 2001; an earlier, coarser, resolution dataset dates to 1976. These datasets allow such phenomena as wetland restoration and urbanization to be analyzed over time.

Land cover data are available through the USDA Geospatial Data Gateway, the USGS Land Cover Institute, and the USGS National Map portal. The USGS also hosts land cover data by state.

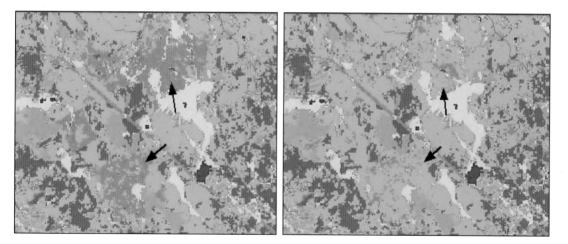

Samples of 1995 (left) and 2001 (right) NLCD 2001 land cover data. Black arrows indicate areas where shrublands have regrown to mixed or deciduous forest in that time period. The location is Kenduskeag and Glenburn in the Kenduskeag Stream watershed of Maine. Courtesy of NOAA.

INVESTIGATE YOUR OWN LAND COVER CHANGE

Access the USGS seamless data server. Zoom in to a small area in a region that interests you and download the 1992 and 2001 National Land Cover Data (NLCD) layers for that region. Unzip the two data layers and add them to ArcMap. Analyze the metadata for the land cover categories and, using the metadata, add and populate a field in each of the land cover tables indicating the land cover type that corresponds to each value. Using Spatial Analyst, calculate statistics for the land cover categories in your region. As each cell covers 30 × 30 meters = 900 square meters of terrain, add a field that indicates the amount of terrain covered.

How much terrain is covered by each land cover type in 1992?
How much terrain is covered by each land cover type in 2001?
Which land cover type has undergone the most change? Why?

Population data

Traditionally available as vector data, population data are also available as a raster dataset called the Gridded Population of the World. The dataset was developed by the Center for International Earth Science Information Network (CIESIN) at Columbia University in collaboration with numerous organizations, including the United Nations Environment Programme (UNEP). A driving motivation behind the project was the desire to address the differences in the sizes of areas used for census taking, the methods of census taking, and the differences in the frequency of census taking. In its third edition, this dataset portrays the spatial distribution of human population across the globe as a series of georeferenced quadrilateral grids at a resolution of 2.5 arc-minutes. The data are available in several formats, including ArcGrid, from Columbia University's SocioEconomic Data and Applications Center (SEDAC). CIESIN also makes available, through its Place I and Place II datasets, spreadsheets of geospatially calculated values of population and land area in specific zones of population density, coastal proximity, climate, elevation, and biomes.

CIESIN and colleagues at the International Food Policy Research Institute, the World Bank, and International Center for Tropical Agriculture (Centro Internacional de Agricultura Tropical) convened to construct a global database of urban and rural areas that could also be used to reallocate population. This effort led to the production of a database now known as the Global Rural-Urban Mapping Project (GRUMP) data products, available on the same website at Columbia University. The population grid produced by GRUMP is lightly modeled using urban areas rather than roads. The dataset includes a gridded population projection to 2015. A related dataset, LandScan, was generated at a resolution of 30 arc seconds × 30 arc seconds for the purpose of estimating populations at risk from human-caused or natural disasters. LandScan is available on the Oak Ridge National Laboratory Geographic Information Science and Technology (GIST) website. It is based on proximity to roads, slope, land cover, night-time lights, and other information. The LandScan dataset is copyrighted, and no third-party distribution of all, or parts, of the data is authorized without the appropriate license. However, the data are available free of charge for US federal government and research use at bona fide educational institutions. Commercial, nonprofit, and personal use license fees are determined on a case-by-case basis. However, all users must register to obtain access to the datasets on the website.

Climate and environmental data

Climate and environmental data come in all shapes and sizes, but given their continuous nature, much of this information is available as raster data. WorldClim is a set of grids (at resolutions down to 1 kilometer) for past, current, and future conditions based on output from the Global Climate Model. Data include temperature, precipitation, and altitude. A unique component of this database are the bioclimate variables (BIOCLIM) derived from temperature and precipitation data and used in ecological niche modeling. This includes such variables as mean diurnal range, isothermality, and precipitation of wettest month, among many others.

Another primary resource for climate data is the Climatic Research Unit at the University of East Anglia in the United Kingdom, which has, since 1982, made available gridded datasets of surface temperature data. These datasets were developed from data acquired from weather stations around the world. Most of these weather stations are run by national government meteorological services, and they exchange these data over the CLIMAT network of the World Meteorological Organization's (WMO) Global Telecommunications System. The original 1980s data came from publications titled "World Weather Records" and from the Global Historical Climatology Network at the National Climatic Data Center. The data are in ASCII grid and NetCDF format, and so easily read by many commercial data-processing packages, such as IDL (Interactive Data Language), and public domain software, such as ncview, a NetCDF viewer, and NCL, a scriptable data-manipulation and visualization package.

The Consultative Group on International Agricultural Research Consortium for Spatial Information (CGIAR-CSI) provides global climate data for the world. These data are provided in 5 degree × 5 degree tiles by continent and are based on a rigorous set of climate models. Another useful climate dataset is CGIAR's Global Potential Evapo-Transpiration (Global-PET) and Global-Aridity indices. They provide high-resolution files related to evapo-transpiration processes and rainfall deficit to model potential vegetative growth. These datasets are based on modeling and analyses by Antonio Trabucco (currently at the forest, ecology, and management research group at Katholieke Universiteit Leuven), with the support of the International Water Management Institute (IWMI) and the International Centre for Integrated Mountain Development (ICIMOD). They are provided online by the Consultative Group on International Agricultural Research's Consortium for Spatial Information (CGIAR-CSI) with the support of the International Center for Tropical Agriculture. These datasets are available for noncommercial use in ArcInfo Grid format, at 30 arc seconds (approximately 1 kilometer at the equator), to support studies contributing to sustainable development, biodiversity, environmental conservation, poverty alleviation, and adaption to climate change globally, particularly in developing countries.

These datasets may be downloaded and include the following layers:

- monthly mean potential evapo-transpiration, the monthly averages over the period 1950–2000, with twelve layers available, one for each month
- mean annual potential evapo-transpiration, the annual average over the period 1950–2000, with one data layer
- global-aridity data layer, the mean annual aridity, the annual average over the period 1950–2000, as one data layer

The site provides a methodology, dataset description, and a noncommercial use policy. The authors of the data "recommend that all users who intend to apply these datasets to assess more regional and local conditions should consider the issues of scale, resolution and degree of interpolation of the original data." The dataset may be freely used for noncommercial scientific and educational purposes, provided it is described as the CGIAR-CSI Global-Aridity and Global-PET Geospatial Database.

The Global Inventory Modeling and Mapping Studies (GIMMS) dataset, hosted by NASA, is a global measure of Normalized Difference Vegetation Index (NDVI) covering several decades. It was originally generated to characterize biophysical change. NDVI is a measure of vegetation used in hundreds of climate and biogeochemical models to calculate photosynthesis, the exchange of CO_2 between the atmosphere and the land surface, land-surface evapo-transpiration, and the absorption and release of energy by the land surface.

The PRISM (Parameter-Elevation Regressions on Independent Slopes Model) climate mapping data is the primary source of national climate data on precipitation and temperature in the United States. The datasets were developed through projects funded partly by NRCS, USDA Forest Service, and the NOAA Office of Global Programs, although there is little funding for updating or expanding the datasets. PRISM is a system that uses point measurements of precipitation, temperature, and other climatic factors to produce continuous, digital grid estimates of monthly, yearly, and event-based climatic parameters. This tool incorporates point data, a DEM, and expert knowledge of complex climatic extremes, including rain shadows, coastal effects, and temperature inversions. PRISM data are provided in ArcInfo Grid format, with monthly precipitation and temperature data updated continuously from 1997, with approximate cell sizes of 4 kilometers × 4 kilometers.

Other climate-related datasets for the United States, the polar regions, and the world are provided by the National Snow and Ice Data Center. These datasets include, among other things, a sea ice index, permafrost, and snow cover, most of which are in a raster format. Climate data in vector form do exist, largely as data from weather stations. For example, the German meteorological agency provides access to data for forty-four weather stations in Germany, as well as data from the Global Precipitation Climatology Center. The Dutch Meteorological online database provides weather station and gridded, interpolated data via the Royal Netherlands Meteorological Institute (Koninklijk Nederlands Meteorologisch Instituut) web application, KNMI Climate Explorer. NOAA's Global Historical Climatology Network (GHCN) version 2 contains historical temperature, precipitation, and pressure data for thousands of land stations worldwide.

The UN FAO provides up to fourteen parameters as monthly data for 28,100 weather stations in datasets tailored to the user's needs as part of the Climpag project. The Vulcan project from Purdue University is one of a growing number of environmental projects that not only uses GIS, but also makes the spatial data available for others to use. The project, funded by NASA and the US Department of Energy, quantifies North American fossil fuel carbon dioxide emissions. The project provides raster data on US fossil fuel carbon dioxide emissions in NetCDF (a machine-independent data format), readable by ArcGIS. As climate is important to agriculture, the USDA's Foreign Agricultural Service's Crop Explorer provides some climate data as well. NOAA's Online

Climate Data provides access to data from thousands of weather stations worldwide; the data tables may be saved in a GIS-compatible format.

A Mapping the Biosphere portal from the Center for Sustainability and the Global Environment (SAGE) at the University of Wisconsin–Madison offers a comprehensive world environmental database. The site offers grid data for download, as well as images for viewing in four categories:

- human impacts (population, literacy, built-up land, growth rate)
- land use (crop lands, urban lands, soil pH)
- ecosystems (precipitation, humidity, evapo-transpiration)
- water resources (runoff, irrigated lands, snow depth)

The schematics section offers data on resource flows and pools that make up individual earth systems, such as the radiation budget, carbon cycle, ozone, and the water cycle.

It is worth noting that some climate information is downloadable as vector data, typically as points from weather stations. One example is the Norwegian-based Rimfrost project, a free public domain resource that contains temperature measurements from more than 500 observation stations worldwide. From both a national and a global perspective, the measurements, when statistically corrected, provide the most reliable estimates of global temperature change. Some of the chosen stations have been operating for hundreds of years, as in the case of the Uppsala station in Sweden, which has provided data since 1720. The Rimfrost data are available in several different languages.

Ocean data

For most of the recorded history of world mapmaking, oceans were either used as a canvas for fanciful monsters, as a convenient place for scale bars, notes, and keys, or simply as an expanse of blue. The twentieth century saw an increased awareness of the importance of oceans in the world's biodiversity, climate, and economy (for example, fisheries and transportation). GIS practitioners increasingly required access to much more detailed information than was previously available.

One outcome of the renewed interest in oceans as a source of geospatial data is the UNESCO Intergovernmental Oceanographic Commission ocean data portal called the International Oceanographic Data and Information Exchange (IODE). Its data holdings include ocean chemistry, currents, weather, and much more, some of which is in GIS-ready format. Less GIS-friendly but still useful for the North Atlantic is the

Many historical maps depicted dragons or other creatures living on land or in the oceans. Courtesy of National Library of Australia.

International Council for the Exploration of the Sea (ICES) dataset for the North Atlantic Ocean.

Bathymetric data, showing the terrain on the bottoms of lakes, seas, and oceans, exists for specific regions of the world, such as the Lake Tahoe data hosted by the USGS. On an ocean scale, scientists working with eleven institutions in eight countries have developed a bathymetric database north of 64 degrees latitude, called the International Bathymetric Chart of the Arctic Ocean, available at a resolution of 2 kilometers from the NOAA website. As the ice thins across the northern polar regions, this dataset becomes particularly important.

The UK Hydrographic Office offers bathymetric data for United Kingdom waters and beyond. It is in the public domain and available for a fee. In terms of a global dataset, Miller, Smith, Kuhn, and Sandwell created a new map of sea floor topography, the Interactive Global Map of Sea Floor Topography Based on Satellite Altimetry and Ship Depth Soundings. This dataset is available online and may be downloaded in tiles. The ETOP01 data from NOAA integrates land topography and ocean bathymetry and is available online or on CD. The Scripps Institution of Oceanography at the University of California, San Diego also offers a variety of downloadable bathymetric datasets.

The challenge with much bathymetric data is that the standard grid and ASCII files generated by most bathymetric software are unreadable in ArcGIS without some pre-processing. This typically involves the use of Generic Mapping Tools (GMT), Unix command-line software that can extract and export bathymetric data. One data portal that bypasses these steps is the General Bathymetric Chart of the Oceans (GEBCO) from the British Oceanographic Data Centre, which operates under the joint auspices of the Intergovernmental Oceanographic Commission of UNESCO and the International Hydrographic Organization. One of the GEBCO formats is an ArcGIS-compatible ASCII export file, available in grids of resolution of 30 seconds and 1 minute.

Sea life data are available as raster data (densities) and vector data (observations from ships or dives), sometimes in the form of real-time, near-real-time, or archived feeds from sensors on the animals themselves. One of the most comprehensive datasets is the OBIS-SEAMAP (Ocean Biogeographic Information System Spatial Ecological Analysis of Megavertebrate Populations) from Duke University. The data contain sea bird, sea mammal, and sea turtle information on a global scale. They may be browsed online, or downloaded from the site as required. The World Database on Marine Protected Areas and the Marine Environmental Data and Information Network are good examples of global ocean data portals. The Virginia Department of Environmental Quality's Coastal GEMS website is an example of a regional marine spatial data resource. One of the most useful archives is maintained by Dawn Wright in a website called Davey Jones' Locker: Seafloor Mapping/Marine & Coastal GIS.

Privacy

Since its inception during the 1960s, GIS has always been surrounded by concerns over privacy. In *Privacy and Freedom*, Alan F. Westin (1970) defined privacy as "the claim of individuals . . . to determine for themselves when, how, and to what extent information about them is communicated to others." There are various reasons for the privacy concerns surrounding GIS. First, maps have always been one of the richest sources of information, allowing people to see patterns and relationships that may never be discernible in simple text and tabular form. Second, information about individuals before the advent of GIS may have been available as public domain data in paper map form, but accessing those data required a visit to the local county or city assessor or public works department. Storing these maps within a GIS environment makes the same information much more accessible and much better integrated with related and nearby information. However, this is only true if the data are made available; otherwise, access to data is just as difficult as that of a paper map. Accessibility has improved a hundredfold with the migration of GIS to the Internet to the point where not only GIS professionals are accessing the data, but also the general public.

Associated with this trend is the exploding number of interconnected devices that are also geo-enabled, such as mobile phones, cameras, live webcams, and vehicles. In a perhaps surprising revelation, two very small local governments in the United Kingdom—Shetland Islands Council and Corby Borough Council—had more CCTV cameras than the San Francisco Police Department. An increasing number of Internet applications, such as Twitter and Facebook, are becoming spatially enabled, allowing you to track your friends or contacts on a map or satellite image. At the same time, advances in science, such as the combination of biometrics and facial analysis, can identify people from CCTV footage (*Contingency Today*, 2009). Just as the construction of city walls, moats, doors, and windows created barriers and expectations of privacy in medieval villages, geo-enabled technologies are, a millennium later, working to break down these barriers.

Third, the resolution of the spatial information managed within a GIS has steadily improved, allowing people to examine overhead and oblique imagery, personal data, ground photographs, and video to examine individual records and even the individuals themselves. The spatial resolution of the base data doesn't even have to be that high. If an object is captured at a high temporal resolution, such as the movements of an individual via a GPS-enabled device, then that object can be tracked and information can be inferred about actions and behaviors.

As technologies seem to evolve faster than laws, how can privacy laws be applied to GIS when GIS is changing so rapidly? Some consider that the rise of these concerns requires a whole new set of legislation—and even a new term, *location privacy*. An article from 2006 sounded a privacy alarm when a map containing confidential information, such as cases of cancer or an infectious disease, could be de-constructed to identify an actual residence (Curtis et al. 2006). Now, identifying individual residences from spatial data is easier than ever, with a variety of simple-to-use tools, high-resolution satellite imagery, online batch geo-coding tools, and Google Street View, which was launched in May 2007.

The gathering of Street View images has caused privacy concerns in several countries around the world, at times provoking street protests and barricades restricting access for Google's camera-enabled vans. However, the courts have ruled both in favor of the individuals and, at other times, in favor of governments and private businesses. A Pennsylvania couple, who filed a lawsuit claiming that Street View on Google Maps was an invasion of their privacy, lost their case. A photograph showing their home was already on display in the county's Office of Property Assessments website, and they failed to convince the court that their home's value was diminished because of its appearance on Street View. On the other hand, a private Minnesota community near St. Paul, unhappy that images of its streets and homes appeared on Street View, demanded Google remove the images; Google subsequently complied. Despite the lawsuits and protests, Google claims it is legally allowed to photograph on private roads, arguing that "complete privacy" no longer exists in this age of satellite and aerial imagery. However, in an apparent weakening of resolve, Google began blurring some faces and license plates using technologies that became available during 2009.

With hundreds of hours of video and images posted each minute to sites such as YouTube, Panoramio, Picasa, and Flickr, even large companies such as Google have difficulty monitoring and removing content of a personal and objectionable nature. Google employees were taken to trial in Italy for allowing a video of a teenager with autism to be posted online on YouTube. Prosecutors argued that Google, which acquired YouTube in 2006, had neither appropriate content filters or adequate staff to monitor the site, and that Google broke Italian privacy law by not preventing the content from being uploaded without the consent of all parties involved. Interestingly, it was removed in the United States as soon as it was brought to the attention of Google but remained on the Italian site for several months.

Google Latitude, a tool that allows contacts to see each other's location, is just one of an increasing variety of tools linking GIS to mobile devices. This has raised some additional concerns in the United States where mobile phone location data are already available to law enforcement agencies; these new tools could enable the government to track the locations of unsuspecting citizens. At the ISS World conference in 2009, it was revealed that Sprint Nextel, with fifty million customers, had received a total of 8,000,000 law enforcement requests for their customers' GPS location data in the space of one year. The company had to build a special web application to process the requests. The concern is that these types of data requests do not meet the threshold for obtaining a warrant, and service providers who are asked for the information do not work for, or report to, the government.

Although courts have generally ruled that, if a person traveling in public places voluntarily conveys location information, they have also recognized that a person does not automatically make public everything he or she does merely by being in a public place. Legal rights to privacy in the United States began with an article published in 1890 in *Harvard Law Review*, where Warren and Brandeis defined privacy as the "right of the individual to be let alone." This was clarified over the years by the court system, in such cases as misappropriation of one's likeness, intrusion upon a person's seclusion, public disclosure of private facts, and publicity in a false light. Federal acts provide protection of privacy under specific circumstances, such as family education, criminal victimization, and electronic communication. One such act, the 1974 Federal Privacy Act, states that "records on individuals by federal agencies must be for a lawful and necessary purpose." The individual has the right to find out what information is being collected. Much of privacy law has to do with limits on the data that the federal government may gather and publish on private individuals. In the United States, Directive DoD 5240.1-R from 1982 regulates the collection, retention, and dissemination of information on United States persons by United States intelligence agencies. Some argue that such laws do not go far enough, while others counter that they already go too far. Still others maintain that because these directives were passed before the electronic age, they are no longer appropriate and need to be updated. Furthermore, as some privacy law does not apply to the private sector, the rise of private-sector GIS muddies the waters of privacy legislation.

Open record laws further compound privacy issues. These laws specify the circumstances under which the public can access government data, which includes geospatial data.

HANDLING PERSONAL DATA IN THE GIS COMMUNITY

In an early article about privacy in GIS, Onsrud and colleagues (1994) recommended adherence to the international Organization for Economic Cooperation and Development's eight fundamental principles in handling personal data in the GIS community:

Collection limitation principle: There should be limits to the collection of personal information. Collection should be lawful, fair, and with the knowledge and consent of the individual.

Data quality principle: Data should be relevant, accurate, complete, and up-to-date.

Purpose specification principle: The purpose of the information should be stated upon collection, and subsequent uses should be limited to those purposes.

Use limitation principle: There should not be any secondary uses of personal information without the consent of the data subject or by the positive authorization of law.

Security safeguards principle: Personal data should be reasonably protected by the data collector.

Openness principle: Developments, practices, and policies with respect to personal data should follow a general policy of openness.

Individual participation principle: Data subjects should be allowed to determine the existence of data files on themselves and be able to inspect and correct data.

Accountability principle: Data controllers, whether in the public or private sectors, should be held accountable for complying with the guidelines.

Do you think that these eight principles have been adhered to as GIS has evolved since the article was written?

Which of the principles do you believe is most often violated and why?

Which principles will pose the greatest risk of being violated in the future?

These laws vary widely from country to country—and even within countries, by state, and by local government. One of the cases that has had a significant effect on spatial data over the past decade was the 2007 case *California First Amendment Coalition v. Santa Clara County*, which directed the county government to comply with the California Public Records Act and provide its parcel basemap data at the cost of duplication on request. Prior to this lawsuit, the county charged over $100,000 for its public data. The case affirmed that the basemap is a public record that the county could not limit access to the few users who were willing to pay the fee.

More recently, the federal government announced twenty open government initiatives for federal agencies, including previously unreleased records from the Departments of Defense, Transportation, General Services Administration, and the Patent and Trademark Office. In late 2009, the Central Platte Natural Resources District in Nebraska sued the USDA's Farm Service Agency, alleging that the USDA denied the district access to GIS data related to farmland. The resulting court decisions in these cases could make more data available not only to GIS users but also to the general public, since the distinction is seldom made as to the type of user who can access the data. Accompanying the increased availability of data is the increased possibility that data will not be considered private in the legal sense. While it was true in the past that GIS software was required to be able to view geospatial data, that is becoming less so today given recent technological advances, easier-to-use devices, and web-based applications.

Since 2009, it seemed that every few days a story appeared about yet another local, regional, or even national government agency opening up its data to the public, one such example being toronto.ca/open. In addition, an independently developed companion website, DataTO, allows the public to request new datasets and set priority levels by rating the datasets. The City of Toronto even allows mashups and applications to be created with its data to "build a better city" (*GIS Talk* 2009). The datasets include streets, address points, business improvement areas, apartments violating standards, day care centers, and events. The city's understanding of geospatial data was evident when it stated that "it is important for everyone to understand what is in these datasets and to understand the codes."

WHAT IS A MASHUP?

A mashup is a webpage or application that dynamically integrates content or functions from two or more sources. Those sources may reside on the same or different map servers, and may be hosted by different organizations using different protocols (e.g., ArcIMS, Web Map Service [WMS]). The idea took hold in GIS with the appearance of browser-side APIs and web services in 2005. In a mapping mashup, at least part of the content or functions are georeferenced. Mapping mashups include basemaps, operational layers (such as crimes, Twitter tweets, or houses), and tools (such as Find Address). An example is Zillow, which integrates home values, taxes, prices, and school zones.

FOR FURTHER READING

See *Web GIS: Principles and Applications* (Esri Press 2010) by Pinde Fu and Jiulin Sun.

Think about the evolving issues surrounding access to data versus privacy of information and answer these questions:

In your role as a GIS user, do you favor wider or narrower access to geospatial data? Why?

Now considering your role as a private citizen, does your opinion about access to geospatial data change? Why or why not?

How do you see these issues playing out in the future, either coming to a resolution or becoming increasingly complicated?

Summary and looking ahead

In this chapter, we have explored some of the major types of raster spatial data, their limitations and benefits, and how to obtain them. We have considered the privacy implications resulting from the increased availability of spatial data to the GIS user and to the general public.

In the next chapter we will look at the cost of spatial data, including an examination of long-standing debates on whether the data users should be charged a fee for data use, and consider other issues that influence the true cost of those data. We will then discuss key issues that affect the availability, quality, and use of those data.

References

Anderson, James R., Ernest E. Hardy, John T. Roach, and Richard E. Witmer. 1976. "A Land Use and Land Cover Classification for Use With Remote Sensor Data." *USGS Professional Paper 964.* Washington, DC: US Department of the Interior.

"Big Brother Has Got Bags of Potential." 2009. *Contingency Today* (August 24). http://www.contingencytoday.com.

Curtis, Andrew J., Jacqueline W. Mills, and Michael Leitner. 2006. "Spatial Confidentiality and GIS: Re-engineering Mortality Locations from Published Maps About Hurricane Katrina." *International Journal of Health Geographics* 5 (October): 44. http://www.ij-healthgeographics.com/content/5/1/44.

GIS Talk. 2009. "Toronto Data Goes Public—Free GIS Data." http://www.educationgis.com/2009/12/toronto-data-goes-public-free-gis-data.html.

Gonzalez, Rafael C., and Richard E. Woods. 2008. *Digital Image Processing.* Upper Saddle River, NJ: Prentice Hall.

Jensen, John R. 1996. *Introductory Digital Image Processing: A Remote Sensing Perspective.* Upper Saddle River, NJ: Prentice Hall.

"One Billion Square Kilometers of Earth Imagery." 2010. *GIM International* (April 2). http://www.gim-international.com/news/id4537-One_Billion_Square_Kilometers_of_Earth_Imagery.html?utm_source=Newsletter&utm_medium=email&utm_campaign=20100406+GIM.

Onsrud, Harlan J., Jeff P. Johnson, and Xavier Lopez. 1994. "Protecting Personal Privacy in Using Geographic Information Systems." *Photogrammetric Engineering and Remote Sensing* LX(9): 1083–1095.

Thurston, Jeff. 2010. "How Does Satellite Imagery Compare with Aerial Photography?" *V1 Magazine* (April 4). http://www.vector1media.com/dialogue/perspectives/12317-how-does-satellite-imagery-compare-with-aerial-photography.

Tobler, Waldo. 1987. "Measuring Spatial Resolution." Proceedings of Land Resources Information Systems Conference, Beijing, 12–16.

Tobler, Waldo. 1988. "Resolution, Resampling, and All That." In *Building Data Bases for Global Science*, edited by H. Mounsey and R. Tomlinson, 129–137, London: Taylor and Francis.

Westin, Alan F. 1970. *Privacy and Freedom*. London: The Bodley Head, Ltd., 508.

4

SPATIAL DATA: TRUE COST AND LOCAL ACCESS

Introduction

The first three chapters have demonstrated how spatial data from government and private sources, used in conjunction with GIS techniques and technologies, can greatly enhance decision-making and problem-solving processes. However, there is much more to solving problems than downloading, unzipping, and analyzing the spatial data. You must also consider how international and national geospatial organizations, legislation, and standards affect the quality, type, and amount of spatial data that can be accessed. These issues, as well as the cost of spatial data, will be addressed in this chapter:

What is the true cost and value of spatial data?
How can the cost and value of spatial data be measured?
How do the policies determining cost and access ultimately affect the availability, quality, and use of spatial data?

To evaluate all sides of the issue, this chapter will compare the commercial policies of two national mapping agencies and discuss the pros and cons of each policy. Following that, we will look at the policies of two local government organizations and discuss how and why their approaches differ.

Free versus fee

Private companies involved in capturing and producing spatial data must either sell their data or sell the services that are derived from those data to remain economically viable. Some companies produce data themselves, while others, as we have seen, take government-provided data and add value to them. End-users may find this added value to be a more attractive alternative to the raw, or unprocessed, data provided by government sources.

Among the many ways to add value include making spatial data available in a greater variety of map projections, file formats, or by user-defined geographic areas. Data may be supplemented with other related data and some vendors offer discounts for purchasing multiple datasets or for repeat purchases. Vendors may add attribute values to support more attribute queries or add topological relationships to support more effective spatial queries. The data are offered in a variety of media or downloadable more reliably or at a faster rate than via a government site.

These and other enhancements make the data easier to acquire and use. Consequently, users may be willing to pay for data that were originally free. Although users are paying for *value-added* data, this may translate into direct savings if they can release staff resources by minimizing, if not completely bypassing, the data preparation stage. Hundreds of companies add value to government data and then resell those data across a broad range of GIS markets in a service industry valued at US$5 billion annually. These companies are sometimes referred to as *conversion* firms, although they typically do much more than convert data; in many cases they create customized, GIS-ready datasets for particular client requirements.

One example is the GeoCommunity's GIS Data Depot through its portal run by MindSites Group LLC. The GIS Data Depot provides government data such as US Geological Survey (USGS) Digital Line Graphs, US Federal Emergency Management Agency (FEMA) flood data, and US Census Topologically Integrated Geographic Encoding and Referencing (TIGER) files. RMSI Inc., founded in 1992 by Stanford University alumni, provides 2D and 3D map datasets to libraries, universities, and private industry, including land use, digital elevation models, points of interest, and buildings. RMSI's clients include the City of Berkeley, Intermap, the Massachusetts Institute of Technology, National Geographic, and Tele Atlas in the United States; and the Countryside Council for Wales, English Nature, Imperial College, Ordnance Survey, and Sport England in the United Kingdom. Lovell Johns provides Ordnance Survey and other data (health, education, business, travel, and more) to public agencies, businesses, and publishers on an international scale.

Other data vendors target specific markets. OGI (OGinfo.com, LLC) provides geologic, pipeline, parcel, waterway, aerial photographs, and other data primarily to energy companies. Telvent provides weather-related GIS data to help aviation, construction, energy, transportation, and water industries manage weather-related risks. LAND INFO Worldwide Mapping LLC specializes in digital topographic data.

This short list of examples illustrates how many data vendors receive data from, and also provide data to, governments and other private vendors. The data provision network is an increasingly tangled web. However, given the escalating cost of providing data, it makes sense for data providers to work closely with each other and with government agencies. Analytical Surveys, now a division of RAMTeCH, for example, has been providing data to cartographers and GIS users for forty years.

Much of the GIS industry has developed over the years on a foundation of government data. Should those government agencies generating spatial data sell their data products or should spatial data be considered a public good and, therefore, publicly available at no cost? Should spatial data be considered a critical value-added service that a government provides for its citizens, who should pay for that service? Without free or inexpensive government data, the many private companies founded on adding value and reselling those data would go out of business, jobs would be lost, and the entire GIS industry would suffer. Does the government have an obligation to stay out of the business of spatial data and let private companies do this work?

What is the proper role for a government agency? Public agencies provide public goods, but what is this public good? The first economist to develop the theory of a public good, Paul Samuelson (1954), defined it as "[goods] which all enjoy in common in the sense that each individual's consumption of such a good leads to no subtractions from any other individual's consumption of that good." This is known as *nonrivalness*. In addition, public goods are *nonexcludable*, meaning that no one can be effectively excluded from using the good. A public good is generally perceived as something that a government agency is better equipped to provide than a private company, a national military force being the most frequently cited example. Other examples, such as health care, highways, or education, are defended as public goods by some but questioned by others who believe that private entities can provide these services more efficiently and with higher quality. For example, on a local level, private security firms could be used much more frequently than local police forces. In such cases, a public good or a public service is not exclusively provided by a public agency, but may be a mix of public and private bodies, or even completely privatized.

What is the true cost of acquiring and using spatial data? To answer that question, consider the cost to society and the cost to the organizations using it. By definition, a public good involves costs and benefits outside the marketplace and these costs and benefits are only partially reflected in market prices and commercial transactions. Even in countries where federal spatial data are free, agencies providing those data are funded through taxes and other public financing mechanisms. In countries where government agencies make their data available for a fee, those fees—high though they may be—often do not fully reflect the cost of generating those data. The set of contour lines on a single topographic map, for example, were originally derived from hundreds or even thousands of hours spent by a cartographer, or team of cartographers, stereocompiling them from aerial photographs. Those aerial photographs, in turn, had to be planned, flown, captured, photogrammetrically processed, scanned and fed into an analog or digital 3D model. This process required several years and thousands of staff-hours for just a few dozen square kilometers of data.

Tele Atlas and NAVTEQ maintain street data for many parts of the globe. In the United States, the data that they maintain originally came from the ten-year effort by the US Census Bureau to build the TIGER system during the 1980s. By 1990, this government effort had resulted in the first nationwide digital street map in the United States, but broadening consumer access would require an enormous investment by private industry. The data may now be found online (through mapping services such as ArcGIS Online, Yahoo!, MapQuest, and Google) and on almost every handheld computer, mobile phone, and GPS receiver. The small fee that may be paid to access those street data or to purchase the GPS receiver does not in any way match the public and private investment to create the data in the first place, nor do the fees pay for the update of those data.

According to one GIS leader (Aronoff 1989), acquiring and processing spatial data may involve up to 80 percent of a typical GIS project. The time it takes for GIS professionals to acquire and process spatial data not only far outweighs the monetary outlay, but time spent on data acquisition translates into a significant portion of the overall cost of any GIS-based project. If data can be acquired in a format that saves even a fraction of the processing costs, it may be worth an end-using organization paying for another organization to process the data. Purchasing data may actually save money in the long run.

Public good is defined as something that provides benefits to society as well as the individual. It has been estimated that federal investment in public domain spatial data provides billions of dollars in benefits to society in the United States alone, through the work undertaken by higher education, business, the military, and government. This does not include the indirect benefit of improved decision-making at all levels, from local to global, by those who use public domain spatial data. It was reported in *Spatial Information in the New Zealand Economy* that the use and reuse of spatial information was estimated to have added NZ$1.2 billion in productivity-related benefits to the country's economy in 2008 (SKM, Ecological Associates, and ACIL Tasman 2009).

Case study: USGS and OSGB

The USGS and Ordnance Survey of Great Britain (OSGB) are both national mapping agencies, but their policies on data copyright and fees have differed significantly over the years. Logos reproduced courtesy of USGS and OSGB.

The USGS was established in 1879 to formalize the surveying that had previously been commissioned as stand-alone expeditions, beginning with that of Lewis and Clark's Corps of Discovery to the Pacific Ocean. Ordnance Survey of Great Britain (OSGB) can trace its roots back to 1791, when the government of the day feared the French Revolution would spill over to England. The government commissioned a project to map the southern coast of England to identify areas that would be vulnerable to invasion from France.

In the centuries that followed, both the USGS and OSGB became national organizations providing data and supporting users throughout the world. The USGS not only makes maps, but monitors water quality, natural hazards, and biological diversity, and performs other functions as a part of the US Department of the Interior. Its annual budget is around $900 million and it employs about 9,000 people in all fifty states. OSGB employs 1,500 people and has become a £100 million (US$167 million)-a-year operation. Both organizations completed the digitizing of thousands of paper topographic maps during the mid-1990s. Both are leaders in providing digital spatial data, such as land cover, transportation, building footprints, aerial photography, satellite imagery, rivers, landforms, and ecoregions, in raster and vector formats.

However, despite their similarities, the agencies have much different policies for pricing and restricting use of their digital spatial data products. USGS data are available either for free or at a nominal fee to cover media costs (such as a DVD or CD-ROM). The one exception to this general policy over the last thirty years was Landsat data, which traditionally cost the data user about $600 per scene. However, in 2008, the USGS announced that Landsat data would also be free.

By contrast, until recently, OSGB imposed fees to use all of its data. In addition, the use of any data derived from OSGB products by other companies and organizations was also subject to stringent licensing controls. However, in 2008 Cabinet Office Minister Tom Watson established the Power of Information Task Force. The role of the task force was to grapple with the following questions:

How can government further catalyze more beneficial creation and sharing of knowledge, and mutual support, between citizens?

What more can and should be done to improve the way government and its agencies publish and share nonpersonal information?

Are there any further notable information opportunities or shortfalls in sectors outside government that those sectors could work to rectify?

READ AND RESPOND

Read the blog post by Richard Allan, chairman of the Power of Information Task Force titled "Geographic Data that Should be Free (In All Senses of the Word)," in which he contends that certain types of geographic data from Ordnance Survey should be freely available.

Which geographic data are the sets that Allan is advocating be freely available?

Which arguments for and against Allan's position do you find the most convincing? Why?

Which position do you favor, and why?

Which type of policy do you think local and national data providers will follow in the future? Why?

Following a subsequent government review in 2010, OSGB has made available at no cost, and with significantly fewer restrictions, a large number of medium- and small-scale vector and raster datasets. The large-scale datasets, including the OSBG flagship product, OS MasterMap, are unaffected by the changes; they remain available at cost and are subject to licensing restrictions. We will discuss this further in chapter 5 when we look at national data portals.

At first glance, for those seeking to make use of these data sources, it may seem that the USGS model is more appropriate because the data are free. However, in light of our discussion about public goods, would you consider digital spatial data a public good? Spatial data benefit almost every industry and activity in society, from locating businesses, to managing agriculture, coordinating emergency responses, ensuring that the thousands of airplanes are routed safely to their destinations, and in many other critical applications. Spatial data are an integral part of the GIS framework that supports and enhances resource management and decision making on a daily basis. This, in turn, saves lives and property and makes a significant contribution to the economy and society as a whole.

Although it seems an unlikely comparison, spatial data and the humble elevator have something in common. The elevator, along with the supporting grid of beams and columns, paved the way for skyscrapers to be built. Inside an elevator, the occupant seldom thinks of how and why it works: It is simply taken for granted that it just does. Similarly, spatial data support a much more holistic approach in everyday decision making, but the data remain behind the scenes as an often under-appreciated silent resource. Spatial data seldom make the news when topsoil is saved due to more efficient contour plowing or when a city's investment in GIS and spatial data saves millions of dollars due to more efficient street repair scheduling. During the response to natural disasters, for example when people were successfully evacuated and others were successfully rescued after Hurricane Irene struck the east coast of North America in 2011, did you ever hear of spatial data getting any credit?

Four main issues arise from the comparison between the fee and free models. First, producing and updating spatial data are expensive. The first nationwide digital street map, the TIGER system, cost the US Census Bureau an estimated $100 million and 3,000,000 person-hours during the 1980s (US Census Bureau 2011). Second, producing and updating spatial data is a long-term operation, with today's investment unlikely to reap benefits in the short-term for several years. Third, unlike bridges and other tangible construction projects, digital spatial data are largely invisible. The necessary vision and framework for national governments to fund long-term, seemingly intangible, projects such as capturing and maintaining spatial data are often lacking because the benefits and results are difficult to quantify and cannot be easily assessed. Fourth, in the United States, government agencies cannot directly appeal to Congress for funds.

As a result, agencies relying on government backing to produce such data are often underfunded. The decline in funding for the USGS alone, for example, means that much of the digital topographic data are derived from topographic maps that are at least forty years out of date. The spatial data produced by many government agencies are difficult to use, dated, or incomplete. Those agencies lack funds to create better interfaces to the data for end users, or to update the data as often as the user community requires.

By contrast, although Ordnance Survey receives government funds for providing end-users with free medium- and small-scale datasets, it still operates under a cost-recovery structure for its very detailed, large-scale mapping products and related services. Unlike the USGS, OSGB is in a good financial position to keep its data current and expand its data services. Just how the recent change to providing some datasets for free will affect OSGB in the long term remains to be seen.

Another difference between the two organizations is in how their spatial data may be used. Copyrights and patents both encourage and inhibit the creation of public goods by providing a legal mechanism to enforce excludability for a limited period of time. Almost all spatial data provided by US government agencies, including the USGS, are not subject to copyright. This approach is tied to the country's historical view of scientific data; if the data were paid for by the citizens through their government, then those data should not be copyrighted or restricted in any way. By contrast, OSGB data are copyrighted and, depending on the dataset involved, are restricted in terms of how often, where, and how they may be used.

As we have discussed before, public goods impose costs on and extend benefits to individuals or organizations and their impact may be only partially reflected in the price that is paid. Digital spatial data have become critically important to society, yet there are huge costs involved in producing and maintaining the data. In the United States, it is often argued that the taxpayers are already paying for the data through the public monies that are allocated to the data-producing agencies. If the public were charged to use the data directly, would they be, in effect, paying twice? In addition, free government data in the United States has spawned a multimillion dollar industry through the emergence of companies that repackage and sell government data. As a result, many more organizations and individuals throughout society have access to spatial data, from the businessperson to the student researcher.

Some argue that the benefits of abundant data not subject to copyright represent one of the best investments by the US government. Some argue that Ordnance Survey copyright restrictions have hindered the use of spatial data and GIS in Britain and, consequently, have cost the country millions of pounds annually as a result of inefficient decision making. Others contend that it is not the trading fund status of Ordnance Survey that has hindered the wider adoption of geospatial information in Britain, but rather the lack of a coordinated government strategy to promote better use of geospatial information. The recently (2008) published report, "Place Matters: The Location Strategy for the United Kingdom," aims to address this issue; we will discuss it further in chapter 5.

The data-user community typically is small in countries where it is considered reasonable to charge fees for data and associated services, and where spatial data providers operate under a cost-recovery mode. Advocates for the free model maintain that its benefits are not limited to GIS users, but extend to the whole of society through better resource management and informed decision making, improved human conditions, and a more sustainable environment.

It is also important to consider the impact on the host organization of selling data versus providing data for free. As mentioned earlier, the USGS sold Landsat data from the 1970s until 2008, and then announced that all Landsat data would be free. During 2008, when Landsat data were commercially available, 19,000 Landsat scenes were sold. By early the following year, 200,000 free Landsat scenes had already been downloaded, two-thirds of which were from faculty, students, and researchers at universities and schools. During the years in which the USGS charged for Landsat, it cost the agency far more to prepare, process, and provide the purchased data than the revenue generated from Landsat sales. In addition, releasing the dozens of staff members providing customer sales support at the USGS EROS Data Center in Sioux Falls meant they could be reassigned to other tasks providing greater benefit to the wider GIS user community.

The free-versus-fee debate is more complex than it first appears. For most people, spatial data are a means to an end in that it's what end-users do with the data in a GIS environment that make the data useful. This viewpoint makes charging fees unjustifiable for some. It also makes it difficult for those who do not pay a fee to fully appreciate the true costs and benefits of spatial data.

Price and copyright model for the geospatial economy

The *Spatial Sustain* blog hosted a debate between Jeff Thurston and Matt Ball of Vector1 Media on access to geospatial data in an the article titled, "What's Best for the Geospatial Economy, free or fee geospatial data?" (*Spatial Sustain* 2008).

Thurston and Ball write and speak internationally about geospatial technology. They are cofounders of Vector1 Media, which provides information to support economical technologies and processes that promote sustainable environments. A summary of the debate follows.

Jeff Thurston: Excellence has a price tag

Thurston observes that the needs of individuals working with geodata and the needs of organizations change as capacity grows and the rate of change increases. He argues that providing access to spatial data for free is a good way to get people and organizations thinking about and working with spatial information, although he concedes that in most cases free data will be of fairly coarse resolution and smaller scale. The more experience we have working with spatial data, Thurston says, the more sophisticated our use of those data becomes and the greater the demand for more accurate and higher-resolution data.

The better the data, the better the information products (such as analytical models, economic forecasting, and environmental predictions) derived from those data. These products become an essential part of government policy and decision making. As our use of spatial information matures, Thurston suggests that modern cities and organizations depend on "geospatial tools and technologies to exist and thrive."

Thurston goes on to contend that geospatial data are as important to modern society as transportation, health services, and education facilities. But how can we expect data to be available for free when all those other services have to be funded? He predicts the geospatial industry "will become much more 'real-time oriented'" with the increasing requirement for continuous and high-resolution data as consumer applications become more complex. Creating and collecting those data are much more expensive than we perhaps realize. The tools to process and visualize those data will also become expensive, says Thurston.

To tackle the many complex issues that face us today, Thurston believes we need more and better geodata and skilled people to develop solutions. He says we also need to develop a system that allows decision makers to sift out good data from bad or useless data. He argues that the question is not whether we should pay for geodata, but rather whether we want to solve the problems of an "economically sustainable world or not?" Solving these problems requires excellent data, technology, and services. Providing spatial data for free undoubtedly encourages countries and organizations to work with spatial information but if we want excellence, we have to pay for it. Therefore, he believes the most important question that the geospatial community needs to ask is "Do you want to be excellent?"

Matt Ball: The value of free data

Ball, on the other hand, observes that without the long history of access to free, public domain federal geospatial data, the geospatial market in the United States wouldn't be as large nor would it be able to support the services it does through the technological innovation many take for granted today. By providing free data, commercial enterprises were able to improve the quality of, and add value to those data, providing a platform for the entire geospatial industry to build upon.

Ball agrees with Thurston that geospatial data are expensive to collect and maintain. However, he notes that by insisting on charging for their data to recover costs, many governments hamper data use and restrict the economic rewards of free market access. In the private sector, the competition to create better data more quickly and cheaply for better sales margins has led to a great deal of innovation in both hardware and software. To illustrate that innovation, Ball cites the success of MapQuest which, in 1996, produced the first online mapping and driving directions website using the US Census Bureau's TIGER files. The company was later bought for US$1.1 billion by America Online, raising millions of dollars for the national coffers in capital gains taxes.

Before the advent of online mapping sites, there was no business model for distributing data for free. However, the Internet and advertising revenue revolutionized that model and many web-based mapping sites emerged. Mapping is a key component of the local search market, with companies willing to spend a great deal of money to ensure their products appear at the top of search lists. To support those local searches, competing companies invest considerable amounts in data and geospatial technologies which, in turn, benefit the companies providing those services. This, Ball refers to as the "economies of free."

Ball claims that long-standing barrier-free access to geospatial information has led to a great many commitments and investments by both commercial and public entities to create systems and solutions for their customers. Although these investments also create enormous value and tax revenue for the federal government, he doubts that anyone in government has adequately analyzed the value of free geospatial data weighed against data-for-a-fee that was consequently used far less.

Ball concludes that free federal data spurred free market competition: "If the data were locked up to begin with, the market would never have taken off. There wouldn't be the level of investment in technology, and we'd be much the poorer in terms of both economic benefit and our knowledge of our world."

Which author—Jeff Thurston or Matt Ball—do you think makes a more convincing argument? Why?

What advantages and disadvantages do you see in selling data versus providing it for free and without copyright?

Do you think the future will bring more pressure on data providers around the world to provide data for free or for a fee? Why?

National spatial data provision policies

While exceptions exist such as the Netherlands, most governments in Europe license the production and use of their spatial data along the lines of the model adopted by Ordnance Survey in Britain. Indeed, most countries throughout the world follow the model of charging for and licensing their data.

As GIS has expanded over the past thirty years into many areas of society, data users are increasingly vocal in their demands for access to quality, current data. This places pressure on government agencies to provide at least part of their data holdings for free. Land Information New Zealand is another example of an agency that until 2000 charged for all data. After 2000, it began running a robust data portal through which it freely distributes a significant amount of its data holdings.

The United Kingdom's Power of Information Task Force's mission was to advise and assist the government on delivering benefit to the public from "new developments in digital media and the use of citizen- and state-generated information in the UK." One example of a new digital media development is a mapping mashup. An example of citizen-generated information is the enormous rise of volunteered geographic information—information about geographic places, features, and networks formerly gathered exclusively by government employees and now gathered by ordinary citizens.

Citizen groups dissatisfied with government spatial data policies and armed with their own mapping technologies have helped improve the options for those seeking public domain data. One example is the Libre Maps Project, whose organizers—all volunteers—wanted a one-stop portal for USGS Digital Raster Graphics and took it upon themselves to create one. Although the data are free, the group encourages donations to support its work. We will return to the topic of these citizen-generated spatial datasets in chapter 8.

The plot thickens: Mashups

The migration to online data services is rapidly becoming a popular way of providing access to spatial data. With the advent of these services, which may incorporate a variety of data sources—public and private, free and for-a-fee—the cost and licensing issues associated with those data are becoming much more complicated.

Ed Parsons, geospatial technologist for Google, comments that Ordnance Survey "was unhappy with local authorities signing up to the Google Maps API terms of service as it required a 'broad' relicensing of the data to Google and the users of Google Maps-based sites." Some of the data in question were originally created by Ordnance Survey as basemaps and updated by local governments.

Local government data

While much spatial data from national government agencies fall under either the fee or free policies, local government data are made available under a variety of arrangements. In the United Kingdom, much of the data collected by local authorities have been captured against a backdrop of large-scale Ordnance Survey data, so-called *derived data*. As such, the data are still subject to Ordnance Survey restrictions as to who may use them and how they may be used. Local authorities are able to publish derived data, so long as doing so supports the service delivery remit of the authority. The derived data may only be made available with the Ordnance Survey product they were derived from and remain subject to Ordnance Survey copyright provisions. In theory, local

READ AND RESPOND

Read Parsons' comments and the ensuing discussion in his blog post, "Who reads the Terms of Service anyway" (*edparsons.com* 2008).

What are the points that Parsons raises with regard to spatial data?

What are some of the chief counterpoints in response?

What do you think will be the final outcome?

Will organizations be able to serve data that may have been derived from licensed sources, or will they be restricted from doing so?

What will be the consequences for society as a result?

Given the dizzying pace mapping technologies are proliferating at present, who will enforce the restrictions that may be established?

Will anyone pay attention to the enforcement?

New technologies are often outlaws in the sense that they usually move faster and become established before the laws that govern, and in some cases restrict, their use come into force, a topic we will return to in chapter 10.

Do you get a sense that the licensing text from Google referred to in Parsons' arguments is open to several interpretations?

Who creates licensing agreement statements?

Do those who write the statements always consult with the data-using community?

Should they consult with the data-using community?

Who should have the final word in interpreting geospatial licensing statements?

Do you think the license agreements adopted by government agencies should be written by those who understand that geospatial data can be considered a public good?

Do you read the terms of service when you are accessing or posting information?

Does your attitude about reading the terms of service change in any way based on this reading or anything else you have read in this chapter?

authorities could make this information available to download but, in practice, the ongoing license fees sought by Ordnance Survey for their large-scale mapping products would make this prohibitively expensive. As such, the data have typically been made available online as static or interactive maps, without the option to download. However, as authorities take advantage of Ordnance Survey's medium- and small-scale datasets that are now free-of-charge and with fewer restrictions on use, those authorities are more at liberty to provide the data to end-users for commercial use and public access.

In the United States, where national and state spatial data are freely available without prohibitive licensing restrictions, the situation within local government is much less uniform. The complexity, range of geographic areas covered, and diversity of organizational size in local government combine to produce a complex framework. This framework includes councils of government where multiple agencies—most commonly a group of cities and counties from one metropolitan area—collaborate on specific services and programs. The spatial data generation, access, and policies are even more diversified at this local level. Some local governments are using new online map service and portal technologies that allow citizen access to their data.

GREELEY'S PORTAL

An example of an online map service using ArcGIS for Server and the Adobe Flex API is from the City of Greeley, in Weld County, Colorado. The city runs a property information portal with its own spatial data, as well as for the data from the Weld County assessors. Greeley plans links to capital improvement projects, city utilities, economic development, and other data.

Tale of two counties

Washington State illustrates the complications of charging for spatial data or providing it for free. This is partly due to the varied amount of funding available for GIS among different state and local jurisdictions. The State of Washington provides GIS data for downloading, free to the public, from individual agency websites, such as the Department of Natural Resources. By contrast, King County and Pierce County (home of the largest cities in the state, Seattle and Tacoma, respectively) not only charge a fee per GIS dataset, but they also provide data services for a fee to smaller cities that need access to county data. These counties use the fees generated as a steady source of revenue for staffing and equipment to maintain the datasets. The counties see this fee-based approach to GIS data as covering the gap between funding received and revenue required. King and Pierce Counties provide data to students at colleges and universities at a much lower fee and also have separate fee schedules for government and private industry. This strategy sets the highest fees for access to the data to the industries that can afford them while still recovering nominal costs from universities and colleges for the data.

However, data policies, as elsewhere in information technology, are in a state of rapid change. In late 2010, King County began to release much of its data holdings for free. The impetus behind the change may have been the advent of portal technologies that finally made it cost effective for county agencies to share data. Coupled with this was the growing awareness that to be fiscally and organizationally efficient, agencies—particularly those in the same governmental unit—should adopt an enterprise GIS approach and work together. Not only could departments that relied on and updated the data be more efficient, but developers could more easily write and share code, making their operations more efficient as well. Legislation was enacted that mandated the development of an open data web portal, requiring all county agencies to publish data to that portal. At the same time, Pierce County required requesters to sign appropriate data disclaimer forms, select a list of desired data layers from the Pierce County site, add them to the requester's "cart," and pay $80 per hour for staff processing time. The data user would then wait one to two weeks for the data to be processed and delivered.

FOR FURTHER INFORMATION

For more information and an explanation of the history of their data provision strategies, visit the Washington State Geospatial Clearinghouse, the King County GIS Center, and Pierce County Information Technology Services websites.

As we have seen in our discussion of legal issues surrounding geospatial data, the local government framework in the United States is further complicated by the requirement to make the data available to citizens under the Freedom of Information Act (FOIA). The interpretation of how this act applies to the provision of spatial data leaves many gray areas.

Compare the spatial data provision arrangements of Jefferson County and Boulder County, two local governments in Colorado. Although they are adjacent geographically and both have long-established GIS departments dating back to the 1980s, they are dissimilar in terms of some of their data access policies. Jefferson County was one of the first counties in the United States to use GIS and hosts the Jeffco Interactive Mapping Application, or jMap. This web-based GIS application includes parcels, election districts, zoning, permits, geologic hazards, aerial photographs, and other layers, and allows the user to query the system, buffer the data, and create reports. The county also hosts an online crime mapping service. Jefferson County, however, does not allow free downloads of its data: The jMap layers may be viewed online or may be ordered for a fee.

Jefferson County's sales policy is listed on its website under "Technical Services," and states,

> "In accordance with state and federal laws, Jefferson County collects certain information about people and property. Some of that information is, by law, confidential, and the county is very careful to never release or divulge it. At the same time, Jefferson County recognizes that there is a legitimate need for companies and individuals to have access to the non-confidential information it collects. In many cases, Jefferson County has added value to the basic information by providing it to the public in easy-to-use formats. Jefferson County hopes that by preparing products for general use, both businesses and individuals are better served. As technology improves and resources are made available, Jefferson County will continue to publish and freely distribute information on its Web site."

The site hosts an impressive array of over 100 layers listed under "GIS products," from historical aerials ($5 per black and white photo print) to land use ($1,000), to parcel data ($4,100).

Boulder County, north of Jefferson County, also offers a web-based GIS service. Although many layers are available for Boulder County, it doesn't quite match the number of data layers offered by Jefferson County. Boulder County also offers certain datasets for sale. The main difference between the two counties in terms of their spatial data access is that Boulder County offers about forty layers for free as downloadable shapefiles. These include agricultural lands, floodplains, contour lines, wetlands, trails, and wildlife habitat. The City of Boulder, a separate government entity, also offers about a dozen layers as downloadable shapefiles. The city also provides electronic maps in PDF format, printed maps for $72 per sheet plus $15 for aerials and contours, $25 for zoning, and $50 for parcels with addresses. About fifteen separate web-based GIS applications are available online for free access.

Summary and looking ahead

This chapter examined some of the issues surrounding the true costs of spatial data. We have looked at several different pricing models and the advantages and disadvantages associated with them. We have seen how spatial data policies affect the availability, quality, and use of those data. We have considered how mashups are changing attitudes towards licensing and the provision of data. We have compared the pricing and licensing policies of the USGS and OSGB and discussed the advantages and disadvantages of each policy. We have also examined the data access policies and data provided by several local governments, including two neighboring local governments in Colorado.

In the next chapter, we will look at some of the national and regional spatial data infrastructures that have emerged and the government and industry spatial data portals and clearinghouses that have been established. We will also discuss the concept of frameworks, define and illustrate why metadata and standards are important for the use and application of spatial data, and why partnerships are becoming increasingly critical in the world of geospatial technologies.

References

Aronoff, Stan. 1989. *Geographic Information Systems: A Management Perspective.* Ottawa: WDL Publications.

Parsons, Ed. 2008. "Who Reads the Terms of Service Anyway." *edparsons.com* (blog). http://www.edparsons.com.

Samuelson, Paul A. 1954. "The Pure Theory of Public Expenditure." *Review of Economics and Statistics* 36 (November) (4): 387–389.

SKM, Ecological Associates, and ACIL Tasman. 2009. "Spatial Information in the New Zealand Economy." *New Zealand Spatial Strategy* (blog). http://www.geospatial.govt.nz/productivityreport/#_ftn1.

Thurston, Jeff. 2008. "What's Best for the Geospatial Economy, Free or Fee Geospatial Data?" *Spatial Sustain* (blog) (February 29). http://www.vector1media.com/ spatialsustain/.

US Census Bureau. 2011. "TIGER Frequently Asked Questions (FAQ)." http://www.census.gov/ geo/www/tiger/faq.txt.

NATIONAL AND STATE DATA PORTALS AND METADATA STANDARDS

Introduction

As you learned in chapter 1, the primary challenge confronting users of geographic information in the early years of GIS was the lack of easily obtainable, and usable, digital spatial data. Although very accurate, detailed, and comprehensive spatial data are still not available for every region of the planet, a greater variety of public domain spatial data exist today than ever before. In this chapter, we will consider the emergence of data portals, both national and state, and examine various spatial data infrastructure (SDI) initiatives worldwide. We will explore the initiatives that have driven the development of data portals, and the impact these portals have on GIS and public domain data users. In chapter 6, we will review some data portals that are more international in focus. Once again, we will not only discuss the spatial data, but perhaps more importantly, encourage you to consider issues and concerns that led to the development of these data portals, and issues that have arisen as a result of their creation.

A myriad of data, formats, and portals

The amount of spatial data available to end users is increasing almost daily. Now, users must deal with data in many different formats, at different levels of accuracy, at different scales, and which were created by different organizations for different purposes and at different times. The data can also be accessed via a myriad of websites in a variety of ways including FTP transfer, map viewers, KML and text-based files for download, web mapping services, streamed data, downloadable maps, images, grids, tables, text files, and real-time spatial data feeds.

The development of geospatial data mirrored the development of other digital data in the information age; data were typically produced in an unstructured environment within an organization or put together by an individual for a particular project. As a result, it is not uncommon for different agencies to take matters into their own hands and start producing similar spatial datasets to meet their own deadlines and organizational requirements, wasting public funds, and duplicating effort. This leads to a proliferation of under-utilized data silos.

While the current availability of spatial data is a welcome development, the sheer volume of data and the often convoluted steps required to access a dataset remain a concern. Despite the availability of broadband connections, the number of data providers, and the various options for providing data, access to data seems to be a growing concern. How can this be? The problem is no longer the lack of data but rather finding the most appropriate data for a particular task. This poses one of the greatest challenges to a GIS analyst today.

It is not, however, the only challenge. The need for data generally outmatches the rate of its production, with an increasingly diverse range of applications demanding access to accurate, current, and easy-to-use data. Compounding the problem has been the lack of metadata collected with the spatial information. As much data were originally produced by a single organization to meet its own requirements, there was little need to document the data's provenance—sources, accuracy, purpose, and so on. As more departments in that same organization, and other outside organizations, accessed the data, the need for accurate and current metadata became increasingly important to those who, although not involved in the original data production, sought to use the data. The increased *availability* of digital spatial data did not necessarily translate into the improved *accessibility* to geographic information. Users are often unaware of the existence of potentially useful data sources, and do not know whether a certain dataset meets their application requirements, or if there are any software or format incompatibilities. Spatial datasets are often scattered, like other information available online, across numerous servers. Some are not well networked and have scant metadata to inform the world they exist. Useful indexes and notices have appeared from time to time, but just as quickly have disappeared or are no longer updated, such as the *free spatial data blog* that was helpful but discontinued when the student who authored it graduated.

Much spatial data are also maintained on storage devices that are not connected to the Internet at all, such as on CD-ROMs or DVDs or on a disk drive on the desk of a GIS analyst or researcher. Even more problematic are spatial data in analog form—a paper aerial photograph, a historical map, a property parcel map, or a sketch of the altitudinal zonation of a mountain range. If all of this data were made more generally available, it could provide a significant additional resource for those seeking to address local and global issues.

However, organizations are seldom funded to provide data to others; if they are, it is generally not their primary business purpose. The staff and infrastructure costs required to convert such data to digital form, collect and maintain the appropriate metadata, and make data available to the wider community, are enormous. Nevertheless, today's improved data transfer capabilities and simple-to-use data management tools make it easier than ever to place spatial data into the public domain. As a result, more and more organizations are doing just that. Unfortunately, because it is now so easy to place data online for the world to see, it is often done in an unstructured manner, compelled by the requirement to "get something out there" quickly. The data are made available without any metadata to inform potential users as to how the data were created and what, if any, restrictions exist governing the use of the data. Many organizations lack the time and resources to complete a thorough requirements survey to identify which datasets would best meet the needs of their own organization or the users of their data. History, it seems, is once again repeating itself.

As a result, over the last twenty years, government agencies, educational institutions, nonprofit organizations, and private industry at the local, regional, national, and international levels have established data portals and metadata standards to address the issues surrounding the production of and access to spatial data. The remainder of this chapter will look at some of those data portals and the standards that have emerged. The following chapter will focus on data portal initiatives and standards at the international level.

The road to a national SDI in the US

During the 1970s, many US government agencies extensively researched and developed computer-assisted cartography. By the 1980s, GIS had become an established technology and was here to stay. Various federal and tribal agencies in the United States including the US Geological Survey (USGS), National Oceanic and Atmospheric Administration (NOAA), the Bureau of Land Management (BLM), the US Forest Service, the National Park Service, the Bureau of Indian Affairs, the US Department of Transportation, the Federal Aviation Administration (FAA), and the US Census Bureau were collecting and providing access to digital spatial data. For many, it was a natural progression from the analog data collection that had been taking place for decades, or in the case of the Census Bureau or the USGS, for over a century.

By the 1990s, state agencies joined forces with federal agencies to provide spatial data to an increasingly wide variety of data users in government, academia, private industry, and nonprofit organizations. As more and more agencies began gathering digital spatial data, issues began to arise that very clearly highlighted the need for coordination, in particular, edge matching and vertical integration. For example, along what had formerly been the edge of a paper map emerged clear differences between the density, or even classification, of features collected on either side of the join. For example, in hydrography data north and south of an edge, streams either disappeared at the former edge or perennials became intermittent, and vice versa. In the days when all maps were paper-based, a common refrain was that an area of interest inevitably coincided with the edge of a map sheet or at the corner where four map sheets adjoined. In the seamless digital environment, merging the sheets into one large dataset did solve the problem of having access to all of the data at the same time, but highlighted the problem of edgematching.

The lack of vertical integration among features meant that contour lines did not change direction at rivers and that elevation data were not "flat" on top of water bodies. These and other related issues often became apparent when working not just with datasets produced by different agencies, but also with different datasets produced by the same agency. Another problem was the lack of consistency between agencies in defining attributes, so a "trail" to the USGS might have been a "four-wheel drive road" to the US Forest Service or an "access path" to the Colorado Department of Natural Resources. As a result, it was difficult to share data among agencies and for each agency to analyze the other's data. Such was the case in a project where agencies based in the United States were working with Instituto Nacional de Estadística y Geografía (INEGI), the national mapping agency of Mexico, to map lands along the entire 3,169-kilometer US-Mexico border. Creating a common data dictionary took many months before seamless mapping could even begin.

To confound things further, it became evident (and in some cases embarrassingly obvious) that some of the very same spatial data were being collected, at the same scale, by two different agencies. This usually resulted from differing agency requirements in terms of attribution, spatial accuracy, or even time frame, where one agency needed the data before another agency was scheduled to finish it. In one example that a coauthor of this book personally witnessed, the Kansas Geological Survey was producing the same geologic unit and fault information as the USGS because the State of Kansas had different time frame and accuracy requirements.

As early as 1980, the US Office of Management and Budget (OMB) sought to identify the scope of digital cartographic activities at the federal level and recommended establishing a centralized database and a schema for building it. In a second study in 1982, the General Accounting Office (now the Government Accountability Office) highlighted a substantial duplication of data gathering in the federal sector. Rather than diminish, this duplication was expected to increase because of a lack of prescribed standards and inadequate interagency coordination. In response, the Federal Interagency Coordinating Committee on Digital Cartography (FICCDC) was established the following year. The committee addressed these issues with representation from the Interior, Agriculture, Commerce, Defense, Energy, Housing and Urban Development, State, and Transportation departments, plus the Federal Emergency Management Agency (FEMA) and NASA. By 1990, these functions were expanded into a new Federal Geographic Data Committee (FGDC), which now coordinates the development, use, sharing, and dissemination of geospatial data on a national basis. It includes nineteen members from the Executive Office of the President, and Cabinet level and independent federal agencies. The Secretary of the Interior chairs the FGDC, with the Deputy Director for Management at OMB the vice chair. Numerous stakeholder organizations also participate in FGDC activities to represent the interests of state and local government, industry, and professional organizations.

The US National Spatial Data Infrastructure (NSDI) initiative was developed in response to the requirement to reuse and share existing data resources. The name *National Spatial Data Infrastructure* was coined by the Mapping Science Committee of the National Research Council (NRC) in its 1993 report on the mandates and responsibilities of the National Mapping Division of the USGS. At the time, this was a groundbreaking concept. Data were now being considered to be infrastructure, something that was just as critical to a successful society as the physical infrastructure of roads and bridges. It marked the emergence of the information age, in which data were considered to be essential for economic growth and prosperity and also for sustainability, sound management, quality education, and a whole host of other societal goods. Just as importantly, data were not considered the only component in this infrastructure, but rather NSDI included "the materials, technology, and people necessary to acquire, process, store, and distribute such information to meet a wider variety of needs" (NRC 1993). The NSDI was considered by the government to be the best approach to foster better intergovernmental relations, to empower state and local governments to develop geographic datasets, and to improve the performance of the federal government.

In April 1994, the NSDI initiative was formally implemented by an executive order of President Bill Clinton. The order outlined a number of federal actions to foster the development of NSDI and reinforced the leadership role of the FGDC. To this day, the FGDC seeks to coordinate the production of a framework of basic digital geographic data from which other data may be derived. It also seeks to implement the National Geospatial Data Clearinghouse and to improve access to geographic data by developing tools for the exchange of data, applications, and results. This will focus efforts into the research and development of architectures and technologies that will facilitate data sharing. The FGDC further seeks to develop educational and training opportunities to increase the awareness and understanding of the benefits of NSDI through the improved collection, management, and use of geographic data.

Since the establishment of the NSDI, federal agencies have put significant effort into the infrastructure through identifying key data elements, assigning responsibilities to develop and maintain spatial data, producing spatial data, and developing and maintaining portals. Although the FGDC is charged with coordinating those efforts, it is not a regulatory agency nor does it have the power to create or enforce rules. It cannot force federal agencies to comply. All agree that coordination is a laudable goal, but priorities and funding issues sometimes reduce coordination to a secondary consideration. Some argue that federal agencies are creating isolated pockets of exemplary resources and portals (stovepipes of excellence) and cooperate only when it suits their own requirements.

Others argue that the NSDI's federal focus often does not meet the needs of tribal, state, or local government, much less the private sector or the general public.

Although the size of the federal government has expanded since 1980, the staffing of agencies responsible for generating NSDI content—whether transportation, geology, hydrography, aerial imagery, or other data—has declined over the same period. Recognizing that stagnant or shrinking budgets and resources could not build the NSDI, federal agencies increasingly turned to partners to help with this undoubtedly enormous task. These partners included private industry, tribal, state, and local governments. One early manifestation of this has been the NSDI Cooperative Agreements Program (CAP), a merit-based funding assistance program that provides seed money to encourage collaborative NSDI resource sharing projects between and among the public and private sector. Since its inception in 2001, this grant program has provided funding for hundreds of projects involving more than 1,000 organizations. Many of those organizations have institutionalized NSDI practices and have become stalwarts of the NSDI, which encourages others to use and become a part of the infrastructure. For example in 2009, the CAP program was responsible for enabling hundreds of datasets, with metadata, to be made available through the Indiana Geological Survey's website, an ArcGIS framework extension was developed to allow users to analyze distributed geospatial framework data, and integrated high resolution (1:4,800) structure and transportation data for West Virginia was integrated into the NSDI. Through CAP partnerships, a potentially dismal future for the NSDI has been turned around, with the result being the continued production and update of spatial data for the end-user.

As we have seen, the FGDC was established as the organization responsible for maintaining the NSDI. The OMB directive prompted efforts to identify key data elements and assign responsibility to maintain and create specific datasets. The FGDC was charged with coordinating those efforts, but curiously the FGDC does not have the power to make or enforce rules. It is staffed by people who are funded by their parent agencies, such as the BLM or USGS, and lacks the authority or support that some feel it needs to make more rapid progress. In addition, although the FGDC was responsible for maintaining and promoting the NSDI, federal agencies have few marketing resources because they are not allowed to lobby or advertise. Some feel promoting the service should be handled by a private firm, taking instruction from the FGDC. Others feel that the NSDI, despite the CAP program, has too much of a federal focus, and does not meet the needs of state, tribal, or local government, nonprofit organizations, or private industry, or even the public for that matter. Some say that individual administrations have done a better job of supporting and maintaining state spatial

READ AND RESPOND

Read Executive Order 12906 (1994), which defined the NSDI, its leadership, the clearinghouse, and data standards activities. Recall that in 1994, most GIS professionals were thinking mainly about their own data and projects or, at most, data and projects for their local department or agency. In that sense, some think that the NSDI was visionary because it specifically stated that spatial data were critical to the nation as an *infrastructure*, just as roads and water lines are considered an infrastructure.

Do you think that the language of the executive order was clear enough?

What could be improved?

What four components does the NSDI include that explain why it is necessary to acquire, process, store, distribute, and improve the utilization of geospatial data?

Why can't just one of the components be in place for it to be successful?

Do you feel that the goals of the executive order have been realized? If not, can you explain why?

Which factors do you think limit the implementation of the NSDI?

Are these the same factors now as those that existed in 1994? If not, can you explain why?

Do national concerns such as security, environmental, and economic issues make the basic tenets of the NSDI more relevant now, or do they present continuing challenges?

What impact do you think that the NSDI has had on the global community?

What impact do you think that the international community will make on the NSDI?

Do you agree with the FGDC's statement that "Current and accurate geospatial data will be readily available to contribute locally, nationally, and globally to economic growth, environmental quality and stability, and social progress"?

data infrastructures, with many state governments creating a geographic information officer (GIO) position to coordinate the work.

Despite its difficulties, efforts like the NSDI have certainly increased *awareness* of the value, use, and management of spatial data within the United States. Internationally, spatial data initiatives have been created and are increasingly being established in many parts of the world. Geoconnections in Canada (formerly known as the Canadian Geospatial Data Infrastructure) and the Australian Spatial Data Infrastructure are just two examples. Let's turn our attention to spatial data infrastructures in the United Kingdom, so we can compare and contrast the situation there with that described for the United States.

READ AND RESPOND

In 2009, the Congressional Research Service published a report titled "Geospatial Information and GIS: Current Issues and Future Challenges" (available on the Federation of American Scientists [FAS] website). Read this report.

How effectively do you feel the FGDC is fulfilling its mission?
How well do you feel that the federal agencies are coordinating with tribal, state, and local agencies in terms of the NSDI?
What are some of the arguments that others have given? Which do you agree with and why?

SDIs and public domain spatial data in the United Kingdom

The situation in the United Kingdom with respect to the development of national and regional data portals is markedly different from that in the United States. As we discussed in chapter 2, central and local government organizations collect, manage, and supply their data against a backdrop of Ordnance Survey base mapping, the majority of which was, until recently, available at a cost and subject to strict licensing controls. This meant those organizations were restricted in how much of their own data, captured against Ordnance Survey data, may be published, who has access to the data, and how the data may be subsequently reused—the derived data dilemma. Consequently very little public domain spatial data have been available in the United Kingdom.

In keeping with other international moves towards a more centralized approach to spatial data management, the National Geospatial Data Framework (NGDF) was launched in the United Kingdom in 1995. Unlike other SDIs, which focused on datasets and databases, NGDF was established to provide a framework for metadata, standards, and improving access to data through better integration of available information. Voluntary representatives from central and local government, academia, and the private sector worked together to establish askGIraffe, a gateway to geospatial information which provided two services:

- Date Locator—a searchable geospatial data catalog
- Data Integrator—providing best practice guidelines for organizations and developing a standard to promote uniformity in the collection and storage of geospatial data

Between 1995 and 2000, the work undertaken by the NGDF provided the foundation for an SDI in the United Kingdom. Following a review in 2000 into its strategic future, the general consensus was NGDF, as a separate entity, had run its course and the best way forward was through merger with the Association for Geographic Information (AGI). In September 2000, NGDF transferred the management and day-to-day operation of askGiraffe to AGI, which relaunched the service as the Gigateway, a national geographic metadata service for the United Kingdom. All organizations involved in creating and managing spatial data were encouraged to contribute to the metadata service, to "increase . . . awareness of and access to geospatial information in the UK." Although little, if any, of the data referenced in the Gigateway site were available in the public domain, the very fact of accessing an inventory of available spatial data proved an invaluable resource. NGDF ceased to exist in 2001.

A number of other recent developments in the United Kingdom will have a profound effect on the availability of spatial data in the public domain and influence the continuing development and direction of the UK's SDI. Much of this has been driven by the European INSPIRE directive (discussed in greater detail in chapter 6) and the statutory requirements for public sector organizations (national, regional, and local) to maintain and share datasets about the environment to support environmental policy and economic development.

In 2008, Ordnance Survey launched the OS Open Space initiative. This made available an API and selected base mapping datasets (including administrative boundaries, 1:250,000 and 1:50,000 color raster, and a 1:50,000 gazetteer) for developers to embed Ordnance Survey data in their web applications. Although the data were not available to download, this did represent a loosening of previously rigorous copyright and licensing restrictions.

During the following year, under the auspices of the Department for Environment, Food, and Rural Affairs (Defra), the UK Location Programme (UKLP) was established to implement the United Kingdom Location Strategy and the European INSPIRE directive. UKLP's objective is to increase access to public sector information and, under the guidance of Sir Tim Berners-Lee, *make public data public* by introducing a new framework (known as the Location Information Infrastructure) for the management of spatial data for the benefit of government, business, and citizens alike.

This framework is based on three basic principles:

- Quantify the available data and avoid duplication.
- Adopt common references.
- Share location-related information through a common infrastructure of standards and technology.

In support of UKLP and the drive to make more data available to the public, a consultation exercise was launched by the UK Government at the end of 2009 to examine the opportunities for making certain Ordnance Survey datasets available for free and without, or with fewer, restrictions on reuse under a licensing arrangement similar to the Creative Commons approach.

CC AND COPYRIGHT LAWS

Recall from chapter 1 that a Creative Commons license works alongside existing copyright laws to help authors and providers license their works freely for certain uses, on certain conditions, or dedicate their works to the public domain.

The government's consultation process ended in March 2010, and on April 1, 2010, the OS OpenData program was launched. OS OpenData represents one of the most significant developments in GI (geographic information) in the United Kingdom in the last twenty-five years. Ordnance Survey, together with the Department of Transport and Defra, will make available at no cost medium- and small-scale vector and raster mapping datasets, including administrative boundary lines, a national road gazetteer, land use statistics, transportation nodes, and postcode locations.

OS FREE DATA

Some of the data products available for free under the new OS OpenData initiative include

- OS Street View—1:10,000 street-level raster mapping
- 1:50 000 Scale Gazetteer—mid-scale gazetteer
- 1:250 000 Scale Colour Raster—small-scale raster mapping
- Boundary-Line—electoral and administrative boundaries
- Meridian 2—mid-scale vector mapping
- Strategi—derived from 1:250,000 small-scale vector mapping
- MiniScale—1:1,000,000 small-scale raster mapping of the whole of Great Britain
- Land-Form PANORAMA—mid-scale vector mapping of contours and spot heights
- OS Locator—national gazetteer of road names
- Code-Point Open—reference points for unique postcode in Great Britain
- OS VectorMap District—mid-scale vector and raster mapping

The full list of all data products available under the new initiative is available on the Ordnance Survey website.

Other supporting initiatives included the launch of the government's new website, data.gov.uk and the London Datastore in January 2010. In addition to providing access to the OS OpenData products, these portals make available large amounts of nonspatial government data—which had previously languished in inaccessible government data stores—in an integrated web-based environment. The data for both OS OpenData and data.gov.uk are made available under the same Open Government License for Public Sector Information agreement. Subject to acknowledging the source and copyright of the data (which remains unaffected by the new agreement) and agreeing to the proper use and representation of the data, users are free to

- copy, distribute and transmit the data;
- adapt the data;
- and exploit the data commercially, through sublicensing, combining with other data or including it in proprietary products and applications.

Ordnance Survey's large scale datasets remain subject to the original licensing and cost arrangements.

To demonstrate the potential of OS OpenData, Ordnance Survey has itself created a series of *fraud maps* that match fraud statistics to their geographic locations.

Organizations such as banks and insurance companies now have free access to spatial data, which they can use in conjunction with their own data to map any given phenomena, such as flooding, house property prices, and car theft throughout Great Britain.

Some data users have bemoaned the fact that not all data that the Ordnance Survey produces and maintains are released in an open format, such as the large-scale (1:1,250 and 1:2,500) OS MasterMap data, which include vector representations of buildings and land parcels for Great Britain. Nonetheless, the release of the smaller-scale data in 2010 will provide excellent opportunities for developers to create products without the restrictive licensing policies and financial overheads that had existed. It remains to be seen exactly what impact this will have on the national SDI and the wider adoption of GI technologies through the UK economy, but the early signs are very encouraging. Within a week of the launch of OS OpenData, the UK Postcodes site, developed around the OS Code-Point Open dataset, was providing public access to a previously restricted dataset.

One thing is clear, spatial data users were quick to take advantage of the new policies and the availability of these new datasets. In the first four weeks after Ordnance Survey launched the OpenData program, users had downloaded approximately four times as much data as had previously been downloaded in one year. OS OpenData should prove to be the catalyst which started a spatial data "chain reaction" within the United Kingdom. By making their data products available for free and with fewer restrictions, Ordnance Survey will liberate the reserves of spatial data collected and maintained by companies and organizations that can now make their data available for commercial use and public access. We will return to this in chapter 7 when we look at some examples of public domain spatial data being put to work.

NEW METADATA SERVICE

Another outcome of the work undertaken by UKLP will be a reworked metadata service to replace the decommissioned Gigateway service. The new service supports the work of the UK location strategy and the INSPIRE directive.

Regional SDI initiatives in the United Kingdom

In addition to the development of an SDI for the UK, there are a number of regional SDI initiatives in England, Wales, Scotland, and Northern Ireland.

England and Wales

Prior to 2011, two national gazetteers, the National Land and Property Gazetteer (NLPG) and the National Street Gazetteer (NSG), provided the definitive source of national address and national street information within England and Wales. The NLPG provides a repository of national address information, collected and maintained by 348 local authorities, for local government throughout England and Wales. Registered, fee-paying users have access to download and distribute the data.

The NSG provides access to the definitive dataset of streets within England and Wales, for street works, highway maintenance, and traffic management, with a total of 172 highway authorities contributing to the gazetteer. The street data are also available to prescribed organizations and users only; if the intended use of the data is considered appropriate, applicants will be registered as a certified data user. The gazetteer was used in planning the Olympic Route Network (ORN) and Paralympic Route Network (PRN), the designated transport routes for the London 2012 Olympic and Paralympic games (*GIM International* 2010).

The contributing organizations for both gazetteers provide regular updates to the two data hubs; from there the data may be accessed by any registered organization that requires the data to support service provision.

However, one significant drawback of the NLPG address database was the inability to link directly to Ordnance Survey address mapping products. In 2010, following a governmental review, the GeoPlace joint venture was announced. This initiative will involve local governments in England and Wales and Ordnance Survey in a groundbreaking, and many say long overdue, project to create a National Address Gazetteer. GeoPlace will maintain the definitive, single address and street database for England and Wales, combining local government address and street gazetteers, the NLPG, the NSG and, significantly, Ordnance Survey's OS MasterMap Address Layer 2 product. The new gazetteer will be free for all government agencies under a Public Sector Mapping Agreement (PSMA), which provides access to Ordnance Survey data for all public sector organizations in England and Wales and is available under license for all commercial users.

Wales

In 2003, the "GI Strategy—Action Plan for Wales" was published by AGI Cymru (Association of Geographic Information Wales). This publication was the result of consultation between AGI Cymru and the Minister for Economic Development at the National Assembly of Wales. The main aims of the Welsh strategy were to develop the strategy with and for the GI community, have Ministerial approval, and promote best practice for the GI community. In 2009, a follow-up report, "Location Wales," reviewed the impact of the INSPIRE directive on the Welsh GI Community and what compliance would mean, and what was still required to deliver the original strategy. The report set out a number of objectives for the GI community in Wales, that consisted of

- quantifying the current data holdings and avoid duplicating them;
- using common reference data;
- adopting standards to facilitate data sharing;
- developing the necessary skills to make the most of location information; and
- improving service delivery.

The various organizations involved, public and private sector, academia and charities and voluntary organizations, are individually responsible for meeting those objectives, with progress monitored by AGI Cymru on an annual basis.

Scotland

In Scotland, responsibility for implementing an SDI resides with the Scottish Parliament. In 2005, the government introduced "One Scotland, One Geography; A Geographic Information Strategy for Scotland."

The One Scotland, One Geography initiative provides a framework for data sharing in the public sector and coordinates the activities of both local and central government departments to develop the data, systems, communication, and infrastructure to support this. As with the rest of the United Kingdom, much of this work is influenced by the requirements of the European INSPIRE program. The framework will consist of

- a model that describes Scotland's SDI;
- a series of supplementary cookbooks (documentation and toolkits) to promote best practice in the development and delivery of web-based spatial data services in Scotland; and
- an INSPIRE-compliant metadata discovery service.

SCOTTISH METADATA

A pilot metadata discovery catalog has now been launched by the Scottish government to help promote the SDI and to demonstrate how geographic information may be shared.

The Scottish SDI Metadata Discovery Catalogue has been developed using the free GeoNetwork open source web-based catalog application for managing spatially referenced resources on the web.

Northern Ireland

Under the direction of the Northern Ireland Executive and following the publication of "GI Strategy for Northern Ireland" in 2003, GeoHub NI, an SDI and spatial data warehouse for both the public and private sector in Northern Ireland, was established in 2008. Members of the public, businesses, and governments alike may register to use the service and view and interrogate data online. Subject to permission from individual data providers, certain datasets are also available to download. The majority of the base mapping data are provided by Ordnance Survey of Northern Ireland (OSNI) and as with OSGB, are subject to licensing and reuse restrictions.

Other national data initiatives

Why should spatial data infrastructures such as the NSDI and the SDIs in the United Kingdom matter to users of public domain spatial data? As we have seen, support from the highest levels of government, accurate and high quality metadata, consistently applied data standards, clear roles for data providers to reduce duplication, and partnerships to produce and update critical spatial data are areas that spatial infrastructure initiatives seek to influence and improve. These are issues that have a critical impact on the work of every GIS user because they affect the quality, utility, and amount of data available. They also impact how much of those data will end up in the public domain. One of the practical ways that spatial data infrastructures can benefit GIS users is through the creation and maintenance of spatial data portals. In addition, an often under-valued yet arguably the most important component of a GIS is people. If people involved in GIS are connected in the type of geospatial community that spatial data infrastructures can foster, then the more effective use and implementation of GIS will ensue.

Spain

INSPIRE has been the driving force behind a number of international data portals and infrastructures. Spain's Permanent Commission of National Geographic High Council established the *Infraestructura de Datos Espaciales de España* (Spatial Data Infrastructures of Spain) in 2002, in accordance with the INSPIRE directive. Its goal is to integrate through the Internet, data, metadata, services, and geographic information produced in Spain, within the state, regional, and local government levels, in accordance with their respective legal frameworks. It provides a portal for some datasets, such as land cover (called CORINE) and can support a degree of user input. At the time of writing, it is possible to integrate user-generated maps and services that are aligned with the Web Mapping Services standard from Open Geographic Consortium and geographic metadata created that conforms to the Metadata Spanish Core and the ISO standard 19115.

The Netherlands

In the Netherlands, the Dutch used the ArcGIS for Server Geoportal extension to set up their own data portal, the Geospatial Data Service Center (GDSC). They also developed their own metadata profiles because they felt the international profiles were not strict enough. The tools they developed were marketed as a metadata editor called Geosticker (Put 2010). Also developed in the Netherlands, the *Nationaal Geo Register* (NGR) indexing service is based on open standards and can be used with ArcGIS. From a geoportal built with the ArcGIS for Server Geoportal extension, you can interrogate the NGR for data with a query based on the amount of precision, the theme, or the spatial location desired. Incorporated into the portal is the *Publieke Dienstverlening op de Kaart* (PDOK), or "public services put on a map." This aims to make geo-information easily accessible for all users and also allow user content, such as geotagged photographs, to be uploaded. This recent phenomenon of two-way exchange of information is becoming increasingly popular and will be further discussed in chapter 9 when we look at volunteered geographic information (VGI).

Slovenia

Slovenia has operated a web-based, publicly available geospatial portal since 2004. The development of this portal is in large part due to the efforts of the Surveying and Mapping Authority of the Republic of Slovenia, which in the early 1970s transformed the entire set of land cadastral data into digital form. This portal offers geographic, mapping, geodetic, land cadastre, building cadastre, house numbers, public utility infrastructure data, real estate market transactions, and much more. Searches may be based on address, land parcel number, and through a number of other queries. This effort spawned interest in geospatial data from other government organizations in the country and led to the development of the very detailed *Atlas Okolja* (Atlas of Environment), the Ministry of Agriculture, Forestry and Food's Graphical Records of Agricultural Units, which includes data on land use and annually-updated agricultural use, and the Digital Encyclopedia of Slovenian Natural and Cultural Heritage.

Croatia

Neighboring Croatia also simplified access to countrywide geographic data through its online geoportal, the first one of its kind in southeastern Europe. Launched in early 2010, GeoPortal DGU is hosted by Croatia's State Geodetic Authority (SGA). The site supports the discovery of national data as well as provides strict access control and data quality policies. The geoportal was developed by Esri distributor GISDATA and Conterra GmbH, the professional services arm of Esri Deutschland (Germany), using the ArcGIS for Server Geoportal extension. Data available through the portal, some of which are for sale, include digital orthophotos, basemaps, administrative units, and land survey information.

The Organized Land Project, which streamlines and regulates the registration of land and property across the country, was one of the first to benefit from having access to the portal. By making data more accessible, the processing time for changes to land titles dropped from an average of 400 days to thirty-seven days. This translates into huge cost and time savings for government agencies and citizens alike.

Despite its many innovations, Croatia's portal is viewed as only the first step in establishing a national geoportal. As with many other data portals in Europe, it was developed in response to the INSPIRE directive to share geographic information across Europe. Not only can organizations outside the country access the data, but those within Croatia can use the framework provided by the portal as a guide for producing their own data. The portal provides spatial data and serves as an agent of organizational change. At present, only a few agencies have made their data available on the portal, but its very existence and the advocacy of those hosting it will encourage other agencies to contribute. Could the portal go one step further by encouraging open access to information and more transparent government? A geoportal has the potential power to be not just an agent of change for organizations but political agents of change as well.

India

As we will explore further in the next section, it's important to note that although many national spatial data infrastructures provide the environment to collaborate, set standards, and influence data collection and management, they do not themselves provide data for download. For example, the National Spatial Data Infrastructure Portal of India provides access to information about spatial data that have been developed by several government agencies in India but not, at the time of this writing, the actual data. The site was organized by Esri and the India Ministry of Communication and Information Technology.

Brazil

In some instances, the major spatial data repository for a country is not the same as the national spatial data infrastructure for that country. For example, in Brazil, the *Instituto Brasileiro de Geografia e Estatistica* (IBGE), or the Institute of Geography and Statistics, has for many years hosted spatial data on biomes, land cover, transportation, hydrography, and census on its website. (You will have the opportunity to use some of their data in the activity that accompanies the next chapter.) However, the country's GIS portal, launched in 2010, is located on a different website, hosted by a different agency, and funded by a different initiative, *Infraestructura Nacional de Dados Espaciais* (INDE) or the National Spatial Data Infrastructure. The INDE goes beyond the

Geography and Statistics Agency to integrate geospatial data from many federal agencies of the Brazilian government. It also provides a framework for state and municipal governments to participate, through the establishment standards for data production, storage, sharing, and dissemination. It also provides training for both data producers and users.

Each national spatial data infrastructure represents the results of months, sometimes years, of negotiation and planning and involves a myriad of organizations, some of which have never worked together before. In the case of Brazil, the NSDI came about through its National Commission of Cartography (CONCAR), a part of the Ministry of Planning, Budget, and Management. Several ministries, federal and state institutions, and enterprise associations involved in geospatial information production and use, participated in the development of the NSDI.

Once an NSDI has been established, the work does not end there. In some ways, it is just beginning. For example, in Brazil, INDE's implementation plan is scheduled for ten years, encompassing three cycles to the end of 2020. At the end of cycle I in 2011, portal SIG Brasil is expected to have a minimum infrastructure of hardware, software, and telecommunications, with geospatial data and metadata search engines providing access to data. Due to its expertise in providing online access to data, the IBGE will construct and operate the INDE access portal. Information in the purview of the infrastructure includes cartographic and topographic data representing boundaries, satellite images, orthophotographs, protected areas, urban and rural properties, transportation, demography and population, and soils. INDE's information is grouped into reference data, including geodetics, topographic and registry charts, thematic data, including among others, vegetation, soils, geology, land cover and land use, and value-added data. The latter is added by users, members of the public or private producers, to supplement the reference data.

Spatial data clearinghouses and portals

A geospatial data clearinghouse, designed to facilitate the discovery, evaluation, and dissemination of digital geospatial data, provides a new model of public access to geographic information. Contrary to popular opinion, a clearinghouse is not a central repository or library where these datasets are stored. Rather, it is a collection of organizations, software, and tools that create a detailed catalog of geospatial data. If data clearinghouses are to provide uninterrupted services, the system must be developed around a fault-tolerant server architecture. Just like most other online data, these geospatial datasets do not reside in one location but are stored and managed on distributed data servers. Otherwise, there could be serious consequences in the event of natural or man-made disasters. Such was the case when the building that housed the source of a certain geospatial dataset for New Orleans was destroyed during Hurricane Katrina. The backup for the dataset was held at a location in nearby Baton Rouge, which although sufficiently remote to be considered safe, suffered a power failure, making the backup copy of the data unavailable. A similar situation occurred in Haiti when its GIS center collapsed during the January 2010 earthquake.

For the clearinghouse to function, it needs to provide accurate, timely, and consistent information to advise users about the theme, scale, region, format, and other characteristics of the data. Users may also need to know where the data, and any mirrored versions, reside. None of that is possible without accurate and complete metadata. A combination of technological innovations and operational requirements led to the development of the metadata that make clearinghouses possible. On the technology side, web-based indexing technologies were initially quite limited in their support for searching geospatial data, primarily due to the fact that Hypertext Markup Language (HTML) was designed for literal text search and matching. There was no support for searching by coordinates, lineage, and other numeric values found in geographic data, nor did HTML provide for the search and retrieval of the other data found in geospatial metadata. It was impossible to search for information stored within dynamic databases on web servers. Early spatial data producers were forced to use software packages outside of GIS, even text editors, to populate metadata. This resulted in a wide variety of styles, quality, and amounts of metadata. As additional software was required to access and edit the metadata, the process was a cumbersome chore and only the most diligent of data providers bothered with it.

An early standard for information search and retrieval was the Z39.50 protocol, adopted as a standard by the American National Standards Institute (ANSI) and the International Organization for Standardization (ISO). Designed by the library community to provide search capabilities for bibliographic catalog entries on the Internet, it became common in commercial text, document indexing, and service software (Nebert 1996). As other information service companies adopted it (such as Ameritech, OCLC, and Chemical Abstracts), well-known fields could be queried, providing support for search interoperability among servers operating within the same community but using different server software. The advent of Extensible Markup Language (XML) enabled more efficient encoding and searching for spatial data. Nowadays, search capabilities are encoded into web browsers and servers.

The provision of tools and forms within GIS software made it possible for the metadata to be collected while the producer or user was working with the data. This ensured metadata capture became an integrated and efficient part of the entire data capture process and, as a result, was more widely accepted. ArcGIS, for example, provides several forms and templates for the encoding of metadata, including those of the FGDC, and several ways to read the metadata.

On the operational side, the vast increase in the amount of data now available has forced the GIS community to realize not only that metadata are valuable, but they are essential for twenty-first century GIS work. The increase in the demand for spatial data has heightened awareness of the need for accurate and current metadata. As expectations increase, organizations and individuals are more likely than ever to capture at least a base amount of metadata in all of their projects. Metadata are no longer limited to documenting data and facilitating data discovery, but according to Wayne (2005) are increasingly used to

- manage in-house data resources;
- assess the utility of those resources;
- encourage data accountability;
- establish data liability;
- monitor the status of projects;
- evaluate return on geospatial data investments; and
- provide a common language for data consumers, and increasingly, contributors.

Portals would not have been possible without metadata.

The coordinating US federal agency of the Geospatial Data Clearinghouse is the FGDC, which operates a registry service of conforming geographic data servers. In addition, it also develops prototype software, formats, and standards for metadata, facilitates discussion among clearinghouse participants, and designs and develops training materials. The participants of the Geospatial Data Clearinghouse also include data suppliers interested in promoting the discovery of digital geographic datasets through their standardized metadata. This may involve government agencies, academic and research institutions, commercial data providers, and nonprofit organizations. Through international development grants, FGDC also assisted in the development of clearinghouse services in Brazil and Costa Rica.

The Geospatial Data Clearinghouse is a decentralized, virtual repository of geospatial metadata that is maintained and served by participating data suppliers. In the early days of the clearinghouse, gateway software was required, located either on the Internet as a web-to-Z39.50 protocol gateway or as a client side application written in Java.

The operation of the Geospatial Data Clearinghouse involves both client-side and server-side applications. The term *client-side application* refers to the user interface for the clearinghouse using readily available web browsers and hyperlinks through which users access the clearinghouse and its metadata information. *Server-side applications* support the collection and maintenance of the geospatial metadata. The metadata generally correspond to the smallest identifiable data product (a file) for which metadata are conventionally collected. Users now take it for granted they can search on a variety of fields to find relevant metadata. However, for quite a few years before metadata were widely available, it was extremely difficult to search for and find spatial data. Only after metadata were included (spurred by the efforts of the FGDC and the data users themselves) could the data be categorized, thus ensuring that users could search by product category, theme, keyword, and area of interest.

Data producers and suppliers began to collect and record metadata pertaining to their datasets according to the Content Standard for Digital Geospatial Metadata created and released by the FGDC. The standard provided a single protocol to access multiple metadata collections or geographic databases without the need to redesign the existing database systems. This, in conjunction with the tools for capturing metadata that GIS software companies increasingly built into their products, hastened the collection of metadata. Metadata gathering began at the federal level but gradually made its way into state and local government, as well. At the same time, private vendors, nonprofit organizations, even researchers and developers in universities, began including metadata collection as part of their data creation process.

The National Geospatial Data Clearinghouse developed as a collection of nodes where data are available. For example, the USGS Geospatial Data Clearinghouse is one such node. The USGS node was originally based on the four principal data themes and divisions of the USGS: biological, mapping, geological, and water resources information. Nowadays, the variety of portals available makes the nodes relatively invisible to the user, but many organizations are still funded and structured according to these nodes. During the mid-1990s, digital spatial libraries, such as the Alexandria Digital Library at the University of California, Santa Barbara, showed users that spatial data could in fact be easily archived, categorized, and accessed, but only if the geospatial data community rallied around a common cause of increased efficiency through the use and availability of spatial data.

A clearinghouse is often much more than a site where data can be discovered. It is really the manifestation of data agreements and cooperative efforts taking place behind the scenes involving international, tribal, federal, state, and local agencies, and increasingly, the private sector and nonprofit organizations. The cooperative efforts began at the federal and state levels with the aim of bringing information to every citizen, reducing the cost of government, and meeting public information access laws mandated by federal and state legislatures.

Geospatial one-stops

As we have seen, the number of websites providing access to spatial data have increased dramatically in the past decade and show no signs of abating. On the contrary, websites have increased even more rapidly than datasets, as different libraries, repositories, and search options have sprung up to provide access to data. This is both a benefit and a challenge to GIS practitioners. Even a single dataset produced by a single agency, such as digital orthophotoquads from the USDA, may be accessed through a number of public, private, and nonprofit websites, such as state data portals, the USGS National Map, and the USDA Geospatial Data Gateway. This raises questions about whether the data are all the same on the different sites, or whether some sites contain data that have been updated or otherwise enhanced. Despite the numerous sites providing access to data, many online conversations on listservs (electronic mailing lists) and blogs have a common theme: "Where can I get data for my specific project?" Most GIS professionals maintain their own bookmarks of useful sites to link to the spatial data resources they require. Sometimes these bookmarks are published on various web pages for others to use. However, like all lists, especially those tied to dynamic web content, they are difficult to maintain. Either way, this is not a particularly efficient solution to locating spatial data.

Is there a better solution? Various organizations have attempted to provide a single point, or a *one-stop*, for users to access spatial data. For example, the GeoCommunity's GIS Data Depot, as we have seen, was an attempt to collate a great amount of government vector data, such as digital line graphs and DEMs, and host them in one place. Governments have also started implementing their own single data portals for GIS users. The NSDI provided the impetus for federal agencies to not only collaborate on who produces what data layers, but also to provide access to the data. To make it easier, faster, and cheaper for collaborating agencies and the GIS community to locate and access data, a presidential e-government initiative, Geospatial One-Stop, was launched in 2003. The one-stop concept fits well with the FGDC's tenets of data sharing by eliminating duplication of effort and maximizing limited staffing resources. The result was a Geospatial One-Stop portal, originally on Geodata.gov until 2010, when it migrated to Data.gov. This portal hosts data from dozens of federal agencies in an indexed and searchable format, made possible by the requirement for participating agencies to create metadata according to FGDC standards. The Data.gov portal and backend services host the expected base layers, but also some data that may be unique to its site, such as historical DRGs produced for New Orleans after Hurricane Katrina.

Does the Data.gov portal truly provide a one-stop search and locate experience for the user? Like many other portals that were set up with good intentions, unless the site is actively maintained and supported with staff and funding resources, it will quickly become obsolete, only to be bypassed for other portals or in favor of user-conducted searches. Some claim that Data.gov suffers from just such a lack of development and support. This may be due to the fact that Data.gov is an interagency effort, which tends to be difficult to fund and support. Most government funding is tied to single agencies and for prescribed time periods. Creating data portals is often politically perceived in the twenty-first century as good for the people because they are intended to support citizen access and promote government transparency. So, they are created and promoted with great fanfare, but once operational, they are often neglected and eventually forgotten. Data.gov was heavily promoted during the new presidential administration of 2009, even before the spatial data was included, as an example of open and transparent government. Even today, Data.gov is not confined to spatial data, but a great deal of other government information, from legislative reports to statistics. As of early 2011, not much had been heard about the site, but promotion began to increase in the geospatial community by the end of the year. Because of their limitations, one-stop efforts may simply become just another entry in the growing list of data portals, leaving the geospatial portal landscape littered with many stops.

Since 2003, the United Nations has conducted five surveys on transparency in government and e-government (United Nations 2010). Of the all websites that enable citizens around the world to track government stimulus funds, 40 percent contained georeferenced information. "Governments are using geographic information systems to provide information in a more contextualized and attractive manner, while facilitating users' comprehension of the data conveyed"(p. 11). The report points to a lack of interoperability: "Taking the e-government projects that present economic recovery plans in the United States as an example, it becomes evident that most geographic information systems applied by state and local governments are not the same, frequently incompatible, and based on proprietary standards" (p. 14). The top five countries measured by the e-government development index were the Republic of Korea, the United States, Canada, the United Kingdom, and the Netherlands. The ranking by geographic region is Europe, the Americas, Asia (almost the same as the world average), Africa, and Oceania (p. 60).

Thus far, it seems as though there is only a limited relationship between development of e-government and provision of transparency (p. 13).

To address the lack of coordination among US states, the Coalition of Geospatial Organizations (COGO) asked Congress in 2009 to establish federal subcommittees with jurisdiction over federal geospatial activities. This was to "make certain that Congress has an effective structure for oversight and legislation over the increasing federal government activity in geospatial technologies, and its relationship with state, regional, local, and tribal government, universities, and the private sector"(*American Surveyor* 2009). The reason cited was that the responsibility for oversight and authorization of federal geospatial activities is spread among more than thirty House and Senate committees and subcommittees, but more than forty federal agencies include geospatial activities as part of their mission and 90 percent of government information contains a geospatial component. This inefficient structure does not contribute to strategic, coordinated policy and investments among the federal agencies. The recommendation was prompted by a 2004 report of the Government Accountability Office (GAO) that found "efforts have not been fully successful in reducing redundancies in geospatial investments" and that "OMB's oversight of federal geospatial activities has not been effective because its methods . . . are insufficiently developed and have not produced consistent and complete information. As a result of these shortcomings, federal agencies are still independently acquiring and maintaining potentially duplicative and costly datasets and systems. Until these problems are resolved, duplicative geospatial investments are likely to persist."

University of California researchers in late 2009 partnered with many universities and government agencies to create DataONE (Data Observation Network for Earth), a global data access and preservation network for earth and environmental scientists. Its focus is to aid in finding, sharing, and supporting environmental research. The collaboration of universities and government agencies, even on an international level, is a trend noted elsewhere in this book, a result of a recognition that no single organization has sufficient influence or resources to coordinate data sharing efforts. Among the activities are research and data acquisition, storage, mining, integration and visualization, which will result in a *cyberinfrastructure* that will be made permanently available for use by the broader international science communities.

It must be stressed that all of these public e-government and collaborative mapping and data initiatives are still in their infancy, and to what extent government should be involved, or stay out of the way, is still being debated. The main benefit of e-government and portal initiatives may not be an ideal end-user experience with the data portal, but the fact that agencies begin to collaborate on geospatial programs, even

DOWNLOAD AND USE DATA FROM SEVERAL ONLINE ONE-STOPS

Choose an area in the United States that interests you or where you will be using GIS in the future. Download a USGS digital orthophoto quadrangle (DOQ) from the National Map website and bring it into ArcGIS.

What about the interface, ease of use, available data, and projection and coordinate system of the data do you like?

What don't you like?

The National Map points to the USGS Seamless Data Server. Try accessing the server directly and downloading data that way, comparing its interface with that of the National Map. Next, access the Microsoft Research Maps website to download a DOQ. Make sure that you also download the registration information as a separate file and name it appropriately; for example, the registration file siouxfalls.jgw must accompany a siouxfalls.jpg topographic map or aerial photograph. Bring it into ArcGIS and make sure it is georegistered by checking the coordinates below the data frame. If it is successfully registered, it will be in UTM and you will see the easting in hundreds of thousands of units, and the northing in millions of units (meters). Use the guidelines contained in the terraserver_procedures.pdf on the Esri GIS Education Community ArcLessons (http://edcommunity.esri.com/arclessons) to help you.

Next, do the same thing by searching for a DOQ of the same area using the Data.gov site. Compare the two sites as to their ease of accessing data and the amount of work you had to do to bring it into ArcGIS. Compare the coordinate systems for each dataset and the work you would have to do to convert them to, say, the North American Albers equal area coordinate system.

Why do you think that multiple sites exist containing the same data?

What are the advantages and disadvantages to multiple sites containing the same data?

What are the advantages and disadvantages of each of the sites?

if the collaboration only lasts for a short time. Spend some time with these portals and make your own assessment.

Consider the discussion about spatial data infrastructures and the portals that allow access to these data resources and reflect on the following:

How do you suppose that spatial data providers decide what datasets belong in these portals?

Even more importantly, who determines what datasets are produced so that duplication is eliminated and users' requirements are met?

These decisions are made through a framework.

Framework

The framework forms the backbone of the data repository in any national spatial data infrastructure. While GIS is used to solve a myriad of issues, problems, and day-to-day work from the local to global scale, GIS users typically have a recurring need for a relatively few themes of data. National data frameworks are often a collaborative community-based effort in which these commonly accessed data themes are developed, maintained, and integrated by public and private organizations within a specific geographic area. In the United States, the framework concept was developed by county, regional, state, federal, tribal, and other organizations under the FGDC to share resources, reduce costs, improve communications, and increase efficiency by reducing duplication.

For the US NSDI, the geospatial data themes include geodetic control, orthoimagery, elevation and bathymetry, transportation, hydrography, cadastre, and governmental units. An example of a portal that houses framework themes is the National Map, hosted by the USGS.

Framework is more than just the data alone, however, because without procedures and tools in place to support access to the data, users would be very restricted in terms of what they could do with the data. Framework also includes institutional relationships and business practices that support the environment of sharing data and is designed to facilitate the production and use of geographic data, to reduce costs, and improve service and decision-making. Technical aspects of the US framework include a feature-based data model, permanent and unique feature identification codes, a reference to modern common horizontal and vertical geodetic datums, seamless data integration for adjacent or overlapping geographic areas, and metadata.

The framework includes several key operational aspects. First, transactional changes are made accessible: Who made a change in the portal, when, and why? Second, access to past versions of the portal is ensured. Past versions are not always easy to find, but the reality is that, just like at the local public library, resources are limited. Not all information can be made available at once. Some information has to be archived or even deleted. Finally, the ability to locate framework data from Data.gov is assured.

Institutional arrangements can aid, but not guarantee, a robust and well-maintained framework. Ideally, the framework data for a geographic area will be developed, maintained, and integrated by the organizations that produce and make use of data for that area. In addition, there is a need to ensure that the framework data can be integrated to support applications for different geographic regions. Progress made over the past decade makes it clear that several aspects are the most critical predictors of a successful framework implementation. The first is executive guidance, the existence of an overseeing body established to provide the vision and direction for framework development. The oversight group cannot exist only on paper; it must be involved in the process—not micromanaging, but keeping the developers moving toward the end goal and avoiding distractions which may serve some specialist requirements but not those of the entire community. The oversight group must include several individuals who are practicing members in the field of GIScience. This makes it more likely that the framework coordinators set realistic goals because they understand the technical and organizational aspects, requirements, and motivations behind those goals.

Other predictors of framework success include the following:

- Data development—creating, maintaining and integrating framework data.
- Coordination—ensuring that the contributions of organizations work together and encouraging productive relationships among participants.
- Data management—ensuring the continued viability of framework data through standards, data security, and disaster recovery.
- Data access—ensuring end users of framework efforts obtain the data; otherwise the entire effort may only benefit the participating organizations and very few, if any, individuals.
- Resource management—estimating income and expenses, obtaining critically needed resources, and providing logistical support, all particularly important for government-sponsored efforts that are prone to short term funding and promotion.
- Monitoring and responding—continually or periodically measuring user satisfaction and providing market analysis to framework coordinators, especially important in our modern-day world of changing user requirements and expectations.

State data portals

Many subnational portals, such as those organized by and for a single state, do an admirable job of meeting the requirements of the data users they serve. Many offer an array of vector and raster data layers for their states, often in a user's choice of projections, in a variety of file formats, via web maps or FTP sites, and with historical and current holdings available. State governments have often gone further than federal agencies in terms of coordinating their state spatial data infrastructures. Somewhat ironically, they have received funding from FGDC to develop the strategic and business plans necessary to implement these efforts. Many states have geographic information officers (GIOs) who coordinate state-level activities and advisory councils that help organize the activities of tribal, municipal, and county governments. State portals are some of the most useful data portals available online today, and almost all of their content consists of public domain spatial data.

One of the oldest, if not the oldest, state data portal is the Texas Natural Resources Information System (TNRIS). TNRIS was founded in 1968, long before most people were thinking about data portals, or even GIS for that matter. TNRIS was mandated by the state, originally as the Texas Water-Oriented Data Bank, and renamed TNRIS in 1972. Its mission has always been to provide a "centralized information system incorporating all Texas natural resource data, socioeconomic data related to natural resources, and indexes related to that data that are collected by state agencies or other entities" (Texas Water Code, 16.021). In its early days, the organization provided access to data on the media of the time—paper and film—including aerial photographs, geologic maps, and topographic maps. Like other state agencies, TNRIS has experienced lean years, but fortunately for its end-users and data consumers, it has always been funded and supported, and helped considerably by its status as a state agency. In many states, providing access to spatial data is only one task—admittedly a large task—for an agency with a much larger mission, such as transportation or water resources. Geospatially focused agencies must continually justify their existence, demonstrate that they support decision making and management for all the other agencies, and show that these benefits make each agency more efficient, and above all, fiscally responsible.

In Montana, the state data portal is the Natural Resource Information System (NRIS). It began, and remains today, coordinated by the state library, a somewhat unusual location but one that makes sense: Librarians know databases and public domain data and understand how to document and provide access to those data. As NRIS is a state agency, responsible for water, land, and other natural resources, it was a good fit to house the state's GIS portal, which provides access to such data as parcels, watersheds, population, soils, and satellite imagery. The Pennsylvania Spatial Data Access (PASDA) site is the official public access geospatial information clearinghouse for that state. It has a node on the NSDI, on the DataOne network, and on Data.gov. This state data portal was developed by Penn State University as a cooperative project of the Governor's Office of Administration, the Office for Information Technology, the Geospatial Technologies Office, and the Penn State Institutes for Energy and the Environment at the university. Funding comes from the Pennsylvania Office for Information Technology and the university.

In other states, however, a state spatial data coordinating organization has yet to become a reality. For example, in Colorado many federal agencies have central or western regional headquarters, a number of international GIS and satellite imagery companies—such as GeoEye and DigitalGlobe—have headquarters, every major university has a GIS program, and its Front Range has been called "GIS Alley." Yet, the state has no single coordinating organization for geospatial technology or a single one-stop data portal. As a result, data users must spend a frustrating amount of time trolling different sites from the state's transportation, wildlife, and natural resources departments, and from state universities to find the spatial data they need. Other states are in the same situation, lacking commitment to an ongoing funding program for a state agency to coordinate dissemination of spatial data. Again, it is challenging to convince elected officials of the long-term benefits provided by a state geospatial coordinating agency or function. That's because elected officials are typically focused on what needs to be accomplished during their current term in office and also because the benefits may be difficult to quantify.

When navigating state data portals, it is important to be aware that many federal agencies also operate portals for each state in which they have landholdings or research projects. These portals may not be directly linked to the federal portal, so you may need to search for them separately. The state portals for the BLM are a good illustration of this. The Arizona portal for the BLM, for example, contains over 100 layers, which include data on historic trails, burro herd management areas, grazing allotments, locatable minerals, and more. Some cover the state, while others cover specific study areas such as the Grand Canyon and Lake Havasu. Investigating these state portals from federal agencies could be time well spent.

To address concerns about the different levels of detail and amount of data in state portals and, more importantly, the data access policies behind the portals, the National States Geographic Information Council (NSGIC) was formed. Originally created by geospatial professionals in state and federal government agencies, it has since expanded to include GIS professionals in nonprofit organizations, industry, and academia. The NSGIC aims to coordinate and support geospatial partnerships and standards, and to achieve this with a technology that is often as "behind the scenes" as GIS is. NSGIC's message to government officials is that efficient and effective government can be achieved through the prudent adoption of geospatial information technologies. In other words, NSGIC seeks to convince state policymakers that support for GIS has an excellent return on investment (ROI), a consideration that is particularly important in these times of severe budgetary restrictions and cuts. NSGIC represents a unified voice in its advocacy on geographic information and technology issues, which provides an advantage over individual states or sets of GIS professionals each sending the same message to state and federal legislatures one at a time. NSGIC promotes prudent geospatial information integration and systems development, reviews legislative and agency actions, promotes positive legislative actions, and provides advice to public and private decision makers. The organization is concerned with creating intelligent maps and databases and enabling public- and private-sector decision makers to make informed and timely decisions. NSGIC, therefore, is involved in the challenges of spatial data standards, availability, copyright, and most other issues that we have been discussing in this book. Not surprisingly, NSGIC supports the concepts and implementation of the NSDI. As a GIS data user, you should keep in touch with NSGIC to understand current issues surrounding public domain spatial data.

Thematic data portals

Many data portals around the world are not organized by state or region; instead they are organized around certain themes. For example, the GeoNetwork of the Food and Agriculture Organization (FAO) of the United Nations, which you used in the exercise for chapter 1, was created to improve access to, and integrated use of, spatial data and information, support decision making, promote multidisciplinary approaches, and enhance the understanding of geographic information benefits. Its theme is sustainable development. DataBasin was created by the Conservation Biology Institute, offering an online system that connects users with spatial datasets, tools, and expertise. Individuals and organizations can explore and download datasets, upload their own data, create and publish projects, form working groups, and produce customized maps that can be easily shared. For fifteen years, the National Biological Information Infrastructure (NBII) Metadata Clearinghouse allowed users to share and access information about research into natural resources, with nearly 100,000 records from over seventy data providers. Its theme, as the name implies, was biological data.

COMPARE GEOSPATIAL EFFORTS AMONG THREE US STATES

Examine three different states—Texas, Oregon, and Ohio—profiled and interviewed by NSGIC.

What do these states have in common in terms of their top three geospatial accomplishments for any given year?

What are the differences?

What differences do you think result from whether the state has a strong geospatial coordinating agency?

What do these states have in common in terms of their top three geospatial goals for the year?

What are the differences, and which differences do you think are due to the condition of the state budget?

What do these states have in common in terms of geospatial challenges?

What do you think the data access issues are in Oregon?

What do you think Ohio means by the turf issues that were lightly touched on?

Which innovative application do you regard as changing the amount and type of spatial data that are accessible, and why?

Contrast the remarks made on these pages between a state like Texas (that has, since the late 1960s, had a funded state agency responsible for geospatial data coordination) with a state like Colorado (which despite being one of the focal points for GIS in the United States, has no such statewide GIS coordinating organization and only relatively recently funded a full time statewide GIS coordinator).

Digital cities

Digital cities are very similar in concept to SDIs in that they integrate data from a variety of sources and disciplines and require high levels of collaboration, interoperability, and new perspectives for governance. They rethink how cities should be run. According to Thurston (2009), they "seek to integrate spatial data around a single, accessible, and multi-purposed model." Currently, much of the effort is in building useful 3D digital city models and porting 2D attribute data to these models; this should lead to enhanced modeling and visualization of urban areas in the future. Digital cities and other new initiatives may initially appear irrelevant to the users' immediate requirements for data access but, as they develop, they will offer data and services.

How do portals and frameworks operate? They are dependent on something that was briefly discussed in chapter 1 and referred to many times in this chapter already—metadata.

Metadata

As you recall from chapter 1, metadata describe the characteristics of a data or information resource. Compared to the data, the metadata file may be small in size, but it is just as important as the data. Some would argue that it is even more important than the spatial data because, without metadata, spatial data have no measure of quality and would remain beyond the reach of geospatial searches. Consider the nutrition facts on a food container that explain the recommended portions, calories, fat, vitamins, protein, and other information about the contents. You don't have to read the information provided to consume the product, but if you are choosing between one food item and another, reading the nutrition facts can help you to make an informed choice. Similarly, when choosing between one dataset and another, the information provided can help sway your decision. Comparable to a library catalog record, a record of geospatial metadata contains a title and an abstract. Metadata are sometimes stored in an XML document, a word-processed file, or even a text file, and nowadays they can be easily edited within GIS software.

Although on a much larger scale and with far-reaching consequences, consider the use of data in a GIS-based project where people's lives, property, habitat, neighborhood, or entire ecosystem hang in the balance. A project's success or failure often hinges on whether the appropriate data are used.

For every news story, the most important elements are the "who, what, when, where, why, and how." In the context of GIS, metadata explain who created the data, what the data contain, when the data were created and updated, where the data were created, why they were created, how they were created and updated, and perhaps how they can be used. Metadata help users find the data they need and determine how best to use the data.

The Content Standard for Digital Geospatial Metadata (CSDGM), which was implemented by the FGDC, recommends that certain items be included in metadata. The FGDC states that "users are strongly encouraged to consider [these items] as a starting place for template development." Interestingly, the FGDC cannot *require* that agencies follow the standard—it is not a regulatory agency—and can only *strongly recommend*, although US federal agencies are, at least on paper, required to follow the standard. However, due to FGDC's extensive efforts over twenty years, together with the growth in the GIS community around the world, most GIS professionals agree, at least in theory, that these metadata elements of who, what, when, and so on, are important. The value of metadata is quickly appreciated when trying to find a particular dataset that is known to exist but can't be located, or when finding an undocumented dataset that may look useful but its provenance can't be confirmed.

Section one of the metadata standard is mandatory and focuses on the identification of the dataset. Identification elements include the originator or creator of the data, the publication date, the title, the purpose of the data, the time period for data contents, the *currentness reference*, the state of the data, the frequency with which the data are maintained and updated, and the bounding coordinates. The theme is identified, along with keywords, and any access or use constraints, along with a point of contact, including the name of the individual and the organization. If the data are available online, the URL to that location is required.

Section two of the standard is mandatory if applicable and focuses on data quality. It includes a logical consistency report, a completeness report, and a process description and date. If source data were used, the originator, date, title, online link, scale, type of source media, time period, currentness, and an indication of contribution must be provided. If data assessments are performed, reports on attribute accuracy, and horizontal and vertical positional accuracy must be included. If imagery is included, cloud cover must be indicated.

Section three of the standard deals with the method under which spatial data are organized. Section four has to do with the spatial reference information, including horizontal datum, ellipsoid name, semi-major axis, denominator of flattening ratio, and the horizontal coordinate system. If the data are in geographic coordinates, the latitude and longitude resolution and the units must be given. If the data are in projected (planar) coordinates, the encoding method must be given, along with the resolution of the abscissa and ordinate and the distance units. If the data are in a map projection, the projection name and parameters must be given. If the data are in a grid coordinate system, the grid name and parameters must be given. Additional elements may be distance and bearing, a local planar or horizontal coordinate system, or a vertical coordinate system.

Section five is concerned with entity and attribute information, the *I* part of GIS. This includes field names and any aliases, the type of data in each field, and any relationships between fields. Section six is concerned with information about how the data can and should be distributed—who the contact is, their organization and position, and contact information. An important part of this section is the distribution liability. The last section, section seven, contains metadata information, including the data, contact, position, organization, the metadata standard name, and the metadata standard version.

Metadata support both data and project management requirements. Metadata support data management by preserving a history of the data so that the data can be reused, reformatted, and adapted, assessing the age and character of an organization's data. That history enables managers to determine which data should be updated, which should be maintained, which should be archived, and which should be deleted. It helps to protect an organization because it limits data liability by explicitly designating the effective and administrative limitations with respect to the use of the data; in effect outlining prohibitions on data use rather than attempting to identify all possible uses. How many times have you clicked through those data liability statements, even in the lessons associated with this book? These statements are a part of metadata.

Another critical component of metadata relates to the attribution of the data. This includes the database element (field, or column) definitions, types, and attribute domain values. Field names are, by necessity, often very short, and it is sometimes difficult to determine what the fields contain. Field name descriptions contained within the metadata provide vital additional information for the dataset user. The quality of the data should also be explicitly stated in terms of spatial and temporal resolution, map accuracy measures, and other standards. However, quality also refers to the overall impression that the metadata provide to the user. Recall our discussion in chapter 2 about fitness for use, placing responsibility on the data user to determine whether a dataset is fit for use. The user can only determine this if adequate truth in labeling exists in the metadata provided by the producer of the data.

Metadata do more than support data management requirements. Metadata help instill a sense of accountability for the data among those who collected and maintained the data. Creating metadata as a matter of organizational policy is much more likely to take root in an organization's culture when all of the employees using data share the responsibility for creating and maintaining the data. This is analogous to public housing projects. Successful projects have all of the residents take responsibility for creating a safe and viable environment. Residents in failed projects believed that everything was the "management's problem." By being forced to declare what they know about their own data, staff members begin to realize not only what they know about those data, but also what they don't know. The embarrassment of not knowing, and realizing that they *should* know something about their own data, is often the driving force to document all datasets.

Metadata also support requirements for project management. Metadata help project planners document the data types and content they require and make the process of estimating costs and scheduling resources more effective. Metadata may also help to monitor the development of data through a regular review of the process steps completed, if those steps are recorded within the metadata. Processes adopted both in-house and by outside contractors can be tracked and metadata can be specified as a "contract deliverable" to those outside parties. Metadata provide all project participants with a common language of attributes and methods, plus a place to record and share their progress.

Geospatial metadata are an essential component for geospatial portals. The FGDC was tasked by Executive Order 12906 to develop procedures and assist in the implementation of a distributed discovery mechanism for national digital geospatial data. The FGDC CSDGM version 1 was adopted in 1994 and updated with version 2 in 1998. The International Standards Organization approved ISO 19115, titled "Metadata for Geographic Information" in 2003. The internationalization of the metadata standards is beneficial because data sharing is supported across national and cultural boundaries. Code lists are enabled to address language and vocabulary differences and improve search capability. (A "trail" in one country's or organization's dataset can be a "road" in another's.) These and other vocabulary and data collection standards can have a huge impact on sharing, searching for, and working on a project involving geospatial data that span multiple countries. We will address international standards in more detail in the next chapter.

Metadata are here to stay, and for public domain spatial data users, this is good news. In the early years of GIS, metadata standards did not exist and metadata were difficult to encode and maintain. Metadata generation required a great deal of time and staff resources, and because GIS users were not as interconnected in a community as they are today, it was difficult for organizations to justify allocating resources to collect and maintain metadata. Nowadays, most GIS users and organizations realize that metadata are critical to the efficient use of spatial data within and between organizations and are an essential component in the field of GIS. Metadata are considered not simply part of the data, but also a best practice within geographic information science. In addition, vetted metadata standards that the community has largely agreed upon and metadata tools inside software such as ArcGIS that support these standards combine to make it easier than ever for users to create metadata. While there are still plenty of undocumented datasets about, particularly with the advent of volunteered geographic information, most core datasets are fairly well documented. However, despite recent advances, collecting and maintaining metadata remains a somewhat laborious task and there doesn't seem to be an easier solution to the problem. Hundreds of articles are published annually proposing new approaches and tools, such as Olfat and colleagues (2010) who proposed a synchronization approach linking metadata to the data to automate the process of keeping the metadata current.

Why are metadata so important to those using GIS and public domain spatial data besides enhancing their ability to locate and use the data they need? Many organizations involved with producing spatial data have been doing so for decades, such as OSGB, the USGS, the United Nations Environment Programme, NASA, and EOSAT. However, departments and individuals within those organizations do change and new organizations unknown a decade ago (such as InterMap, The Nature Conservancy, and Digital Globe) are producing spatial data today. The danger with organizational change is that institutional knowledge is lost as staff members transfer or retire and as methods of creation, storage, and documentation change. New staff may be rightly suspicious of old data that are not well documented; even though the data source may still be valid, and duplication of effort will result if data have to be re-created.

Partnerships for spatial data provision

Despite the fact that the NSDI has been around since 1994, the only accompanying funding was an allocation by the USGS to provide some staff dedicated to the functioning of the FGDC. The federal level made no allocations to develop portals for data that were increasingly in demand from many sectors of society.

Since the early 1970s, the federal government and private sector have spent billions of dollars on satellite-based earth observing systems and have worked with the research community to identify, develop, and distribute real-world applications for mapping, monitoring, and managing natural and environmental resources. Unfortunately, while the existing and potential uses of the technology have been widely cited as valuable to society, and the use of this type of data has greatly increased, the development and distribution of data portals to provide access to the data have remained a challenge for both the federal government and the academic research community.

Satellite imagery and other spatial data are expensive, and using the data requires significant investments in software, hardware, and training. Often, it has been difficult for university researchers to use or even access the data, particularly at smaller schools or research facilities. For thirty years, this has hindered applied research and workforce training. Many state and local agencies working with applied research programs have not been able to effectively integrate remote sensing or GIS technology into their management or decision-support programs.

As the 1990s progressed, spatial data users recognized they could not wait for the federal government alone to act if they were to realize the aims set out by the NSDI. Rather, they needed to come up with an alternate strategy and form innovative and creative partnerships among organizations that had previously not worked closely together. One such partnership emerged in 1998 between the USGS and academic institutions in Ohio. It was named OhioView and began with the capability to view and download Landsat data which, during the 1990s, was still a challenge for many users to freely obtain and examine. OhioView was successful in overcoming some of the major costs besetting the federal government and research communities. Until 2008, Landsat data cost $600 per scene, and slow bandwidth and user-unfriendly formats made them difficult to use; in many cases, reformatting the data required a considerable investment in time and software conversion tools. OhioView's objective was to create a prototype system and infrastructure for high-speed acquisition, processing, and rapid delivery of remotely sensed data to state and local users. The OhioView project remains online today, with thirteen universities participating. The State of Ohio routinely supports the purchase of statewide satellite imagery and enrollment in remote sensing education courses around the state has increased significantly over the past fifteen years.

Data requirements in other states led to AmericaView, modeled after OhioView. Established in 2000, this is a nationwide program focusing on satellite remote sensing data and technologies in support of applied research, K-16 education, workforce development, and technology transfer. AmericaView was formed through a partnership between the USGS and what became the AmericaView Consortium. The AmericaView Consortium is composed of university-led, state-based consortia working together to build a nationwide network of state and local users. AmericaView's goals include an expansion of communications networks, facilities, and capabilities for acquiring and sharing remotely sensed data among consortium members. Beyond the data, America-View Consortium is charged with helping each state overcome challenges with spatial data and helping universities, secondary schools, and the public sector in each state to identify, develop, and distribute the kinds of applications each state needs most. Over a decade after its founding, the program has thirty-six state members and is actively working with the USGS and universities across the country to expand participation in the program to all fifty states.

Events such as the terrorist attacks in 2001, Hurricane Katrina in 2005, and Gulf of Mexico oil spill in 2010 made it clear that achieving a secure and stable digital infrastructure was critical to the United States. However, the AmericaView program, as a nonprofit organization since 2003, has faced continual funding challenges and was reliant on increasingly tight state budgets throughout the 2000s. Despite now including over 350 academic, federal and state agency, commercial sector, and non-profit partners, funding has declined dramatically. However, legislation was introduced to the US Senate in 2009 authorizing a program at the USGS to promote national remote sensing education, outreach, and research. The program also would advance the availability, timely distribution, and widespread use of geospatial imagery for education, research, assessment, and monitoring purposes in each state. If passed, house bill 2489 S 1078, known as the AmericaView Geospatial Imagery Mapping Act, would see AmericaView operate through grant-funded state-level entities known as StateView programs. It would further promote civilian uses of geospatial data at the state and local level through support of education, professional development, training, data archiving, distribution, and applied environmental research. Most importantly, the law would establish a congressionally authorized program to continue this work, rather than the ad-hoc program whose funding has been unpredictable from year to year. The bill recognizes and codifies the value of state-based efforts as integral within the broader USGS mission. Its passage would help ensure adequate funding and significantly enhance efforts, including expanding the program to all fifty states.

EarthScope is another example of an initiative that resulted in a portal focused on a key dataset. This was funded by the National Science Foundation and managed by UNAVCO, an incorporated, tax-free corporation that started off in 1984 as a university consortium, University NAVSTAR Consortium (the group behind the research for the original GPS constellation). Data acquisition, processing, and distribution come from the National Center for Airborne Laser Mapping, Ohio State University, Arizona State University, and the San Diego Supercomputer Center. EarthScope focuses on lidar data, generated by optical remote sensing sensors bouncing laser pulses off objects to detect a variety of phenomena, such as atmospheric constituents, fault lines, and elevation. For many years, lidar data were expensive and difficult to obtain. Now, much lidar data are freely available in several formats from the OpenTopography portal. One format, kmz, shows filtered bare earth and unfiltered hillshade images from two different illumination angles. Another format for more advanced GIS applications, is 0.5-meter resolution DEMs stored as Arc Binary grids, available as 1-square-kilometer tiles. Data can be browsed and downloaded via an interactive map available online. The site, like many others today, provides access to unprocessed data in the form of raw point cloud data. OpenTopography also provides web-based tools to process these data into custom DEM products.

Are these partnerships a natural result of the groundbreaking efforts begun by the NSDI, or have they resulted from the inadequacies of the NSDI? US congressional hearings on the AmericaView Geospatial Imagery Mapping Act recorded not only a review of the NSDI's successes but also its shortcomings. Several witnesses spoke of the need to expand the nation's geospatial imagery and mapping program. For example, in one opening statement, Representative Jim Costa, D-CA, chairman of the Subcommittee on Energy and Mineral Resources, recognized the work of the NSDI and questioned whether a reevaluation of how the federal government manages geospatial data is needed. Karen Siderelis, the geospatial information officer and acting chair for the FGDC, highlighted several achievements of the NSDI, and listed three goals:

- Engaging the nation in a dialog about its geospatial future by holding a National Geospatial Open Forum and using new media to garner input from all corners of the country to seek out the best ideas for enhancing the National Spatial Data Infrastructure.
- Bringing creative energy to making Imagery for the Nation (IFTN) a reality by listening to the nonfederal stakeholders.
- Bolstering the existing geospatial governance structure and ensuring that the FGDC is successful in providing unprecedented leadership to meet the geospatial needs of the federal government and of the nation in the twenty-first century.

John Palatiello, executive director of the Management Association for Private Photogrammetric Surveyors (MAPPS), discussed what he saw as a failure to achieve one of the goals of the NSDI, which was to coordinate federal mapping and geospatial activities by compiling the data "under one roof." The result has been a proliferation of new geospatial initiatives to complete the work that was not accomplished by the NSDI.

Commercial spatial data portals

Given the growing data demands by traditional GIS users, the growing number of nontraditional professionals interested in using spatial data, and the ease of creating online data repositories, the past few years have seen a proliferation of data portals created by private companies. For example, when you considered the best location for siting the fire tower in the Loess Hills of Nebraska in an exercise accompanying this book, you used the GeoCommunity's GIS Data Depot. This was one of the earliest commercial portals providing US federal spatial data in a format and schema that was easier to use than that available via comparable portals at the USGS, US Census Bureau, and elsewhere. Another early arrival was IntraSearch's MapMart, and while most of its data was offered for purchase, samples could and may still be downloaded from the portal.

In 2000, Esri introduced the Geography Network, a service that allowed data users to create maps online using a tool called ArcExplorer Web, to stream in data into desktop GIS software applications, and to download spatial data. In 2008, Esri began porting the Geography Network content to its new online repository, ArcGIS Online. ArcGIS Online originally featured basemaps, such as topographic maps, satellite imagery, and demographic data in an easy-to-stream format that could be imported into ArcGIS for Desktop software. Over the next few years, ArcGIS Online was expanded to include user-generated data sharable with everyone or with specific user groups established by the users themselves. It has now developed into a data sharing community, and in 2010 was relaunched as the ArcGIS website.

In another example of a private company hosting spatial data, MapCruzin states on its website that it is "an independent firm specializing in innovative Geographic Information System (GIS) projects, environmental and socio-demographic research, Web site development and hosting." MapCruzin began in 1996, creating the first US-based interactive toxic chemical facility maps on the Internet and has been online ever since. While MapCruzin's focus is on environmental justice and equality, especially as it pertains to pollution, climate change, peak oil, energy, water, housing, food, and technology, the company hosts a great deal of spatial data, for example, oil and gas leases, geology, climate, and toxic release inventories. Some of the data are from US government agencies and some are from OpenStreetMap and other volunteered geographic information sources. The site, as with other commercial data portals, does host many advertisements but you may find it easier to obtain data here than from government agencies or citizen-collected data sites.

When portals change or are no longer available

The dynamic nature of the web and GIS means that inevitably, portals change or go offline. For example, after fifteen years of operation, the National Biological Information Infrastructure portal was shut down by federal budget cuts announced in 2011. The NBII provided text, pictures, maps, and spatial data about plants and animals. It contained data, tools, and applications from many government agencies, academia, nongovernment organizations, and private industry. Even the loss of this portal is small in comparison to the programs that the NBII supported, including such citizen science initiatives as the Mid-Winter Bald Eagle Survey and the North American Reporting Center for Amphibian Malformations. Fortunately, the NBII metadata clearinghouse was not removed, but rather, moved to the USGS metadata clearinghouse. However, the pointers inside the metadata to the data's location could very well point to locations that no longer exist.

Another example of a closed data portal is the Conservation Geoportal, created to provide a comprehensive listing of GIS datasets and web mapping services relating to biodiversity conservation. Launched in 2006, it grew to over 4,000 GIS metadata entries and allowed users to create live mapping services. It was sponsored by The Nature Conservancy, National Geographic Society, Esri, and the UNEP-World Conservation Monitoring Centre. Unfortunately, in July 2009 the Conservation Geoportal was decommissioned due to a lack of funds required to maintain the site and migrate it to an updated platform and long-term home.

Frank Biasi (2009) of National Geographic Maps posted in an online discussion ten lessons learned from this initiative:

1. Outreach and promotion through various means is essential.
2. Data/metadata publishing should be required and supported by managers and funders.
3. In-kind support is great, but funds for maintenance, upgrades, curation, and marketing are essential.

4. Centralized metadata creation is effective and efficient if funds are available.
5. Portals should allow filtering by organizations, including branded subportals.
6. Portals should support organizations' internal and external publishing needs.
7. Without dedicated stewards, browse "channels" should be populated automatically, not manually.
8. Usability and simplicity in finding and posting content is essential.
9. Map viewers should be simple and usable for nontechnical staff.
10. The concept of sharing data is much more advanced than the practice.

Of all ten lessons, perhaps that last one was the most telling of all.

Lessons learned from a defunct portal

Despite the involvement of long established partners, the Conservation Geoportal had to close just three years after it began.

Why do you suppose this occurred?
What impact do you think its closure had on the user community?
Do you think that other portals are in danger of closing? Why?
Do you think that these lessons apply to the development and maintenance of other geodata portals?
What do you think is the most important of the ten lessons? Why?

Also of concern is the process of overwriting datasets with new content while not saving historical versions. These versions may need to be preserved for their legal, fiscal, analytical, and historical value. A project called GeoMapp, supported by the US Library of Congress, seeks partners to develop best practices for archiving data. The data can be used to model climate change, track land use change, and study demographic trends.

Do you think such efforts will be successful, or will societies and organizations continue to blaze ahead without bothering to archive?
What is lost when old versions of any dataset, particularly mapped data, are lost?

As we noted in chapter 2, the year 2010 also saw the demise of one of the longest serving online spatial datasets, the Digital Chart of the World (DCW). In its heyday, DCW was probably the only source of low-resolution, global vector data. It has been withdrawn now due to the data being inaccurate and out-of-date. Originally hosted by

Penn State University, visitors to the site are now directed to the Natural Earth website. The Natural Earth project was launched in 2008 as collaborative initiative involving a number of volunteers. Some of the data from DCW have been redeveloped into other products, such as the 1:10,000-, 1:50,000-, and 1:110,000-scale glaciated areas vector datasets.

Given the fate suffered by other portals, do you think Natural Earth will last the distance?

Spatial data infrastructures past, present, and future

Spatial data infrastructures are not simply additional partnerships and tools to share spatial data. The concept of an SDI, as with GIS in general, is integrally tied to how an organization operates. Building successful SDIs involves new approaches and flexibility, involving different organizational cultures, knowledge, and ways of doing business. If they appear to be slow moving, this is largely due to the fact that cultures change much more slowly than the technology. They involve many types of organizations, people, data types, and diverse requirements. To be successful, they also require patience; a willingness to learn from experience, limitations, and obstacles that will inevitably be encountered; and the flexibility to be reinvented as efforts proceed. As a result, funneling all of these efforts into a single SDI approach seems unlikely. To date, most SDI efforts are modest and, while continuing to expand, deal mostly with serving existing information, rather than planning for future information. When they near completion, recommendations and wish lists often appear for other kinds of data and applications, as user requirements and available technologies are continually expanding. An SDI's work is never done.

Reflect on these questions about data portals:

Is developing an SDI today a bit like closing the stable door after the horse has bolted?

Is it difficult to rein in partners, to develop an infrastructure, and apply some rules governing the access to data, once those data have enjoyed some freedom?

If organization A can obtain some data its own way more quickly, with less expense, with more control over the data, and with fewer restrictions, why should that organization bother contributing to, or even use, a portal?

Given the ease of placing data online, the variety of tools to do so, and the multitude of organizations and users involved in making data available data, will portals ever become the one-stop shops they aspire to be?

Have we come to the point where portals are no longer needed?

Can data users find everything they need through a standard web search?

Indeed, a Denver GIS consulting firm, The Timoney Group, advertised a workshop in early 2010 titled "Making Spatial Data Google-able." It was targeted at data producers and covered five steps to making spatial data easily findable, viewable, and linkable. The argument was that, given the gathering momentum of open data and open government in the United States and abroad, public organizations are revisiting how best to make their geospatial information available to the public in more user-friendly ways than FTP sites or custom-built mapping portals. The workshop was sponsored by a local chapter of the Urban and Regional Information Systems Association (URISA). Could this be a common workshop and theme of the future? One appeal was that the approach was advertised as equally applicable to enterprise-level systems running ArcGIS for Server, Microsoft's Internet Information Services (IIS) servers, or Oracle-based systems, as well as, at the other end of the scale, a low-cost shared hosting approach, and everything in between. The end goal was to have very simple end-user search instructions and be supported in a standard web browser, perhaps with the instructions for looking up parcel assessment information being a simple as, "Google the address."

However, current web searches, while increasingly able to tap into spatial metadata, are still unable to perform some of the searches possible on a portal. For example, the Data.gov portal referred to earlier supports keywords, spatial coordinates, scales, and allows formats to be specified. There are other portals now providing similar capabilities.

The disadvantage is that every portal has its own search tool, and users must become proficient with each of these tools, as well as visit a number of portals, to find the data they need.

Summary and looking ahead

In this chapter, we have looked at some of the history behind, and current development of, spatial data portals, metadata, standards and the coordination effort associated with spatial data infrastructures at the national level. In a number of countries, we have seen that efforts to establish a national spatial data infrastructure have resulted in several spatial data one-stops rather than one national portal. Despite what some consider to be slow progress, organizational and funding constraints, and an increasingly complex spatial data landscape, we have seen that much progress has been made in the variety and amount of spatial data available through these initiatives. We have also found that these coordinated efforts have resulted in increased efficiency, both in terms of access to and in terms of the use of spatial data.

In the next chapter, we will look at international initiatives that are responsible for promoting access to public domain spatial data.

References

Biasi, Frank. 2009. "Conservation GeoPortal Status." *PPgis.net Open Forum on Participatory Geographic Information Systems and Technologies.* July 1. http://dgroups.org/ViewDiscussion. aspx?c=98db21ba-861e-47bc-8491-ff972bb0b66d&i=66335b72-c19e -486c-a742-b560fde836d5.

"COGO Urges Congress to Establish Geospatial Subcommittee in House and Senate" 2009. *American Surveyor* 6 (January).(1). http://www.amerisurv.com/content/view /5684/121/.

Executive Order no. 12906. 1994. "Coordinating Geographic Data Acquisition and Access: The National Spatial Data Infrastructure." *Federal Register* 59 (April 13), no. 71. http://www. archives.gov/federal-register/executive-orders/pdf/12906.pdf

Nebert, Doug. 1996. "Supporting Search for Spatial Data on the Internet: What It Means to Be a Clearinghouse Node." Proceedings of the 1996 Esri International User Conference, Palm Springs, CA, May 20–24. http://proceedings.esri.com/library/userconf/proc96/t0100/pap096/ p96.htm.

Olfat, Hamed, Abbas Rajabifard, and Mohsen Kalantari. 2010. "A Synchronisation Approach to Automate Spatial Metadata Updating Process." *Coordinates* 6 (March) (3). http://mycoordinates. org/a-synchronisation-approach-to-automate-spatial-metadata -updating-process/.

"Open Roads for London Olympics." 2010 *GIM International* (September 28). http://www.gim-international.com.

Put, Sabine. 2010. "Key elements of Dutch SDI." *Geoconnexion* 9 (July/August) (4): 46–47. http://www.geoconnexion.com/uploads/dutch-sdi_intv9i7.pdf.

Thurston, Jeff. 2009. "How Can Digital Cities and Spatial Data Infrastructure Connect?" V1 Magazine (September 26). http://www.vector1media.com/vectorone/?p=3817.

United Nations. 2010. "United Nations E-Government Survey 2010: Leveraging E-Government at a Time of Financial and Economic Crisis." http://www2.unpan.org/egovkb/documents/2010/E_Gov_2010_Complete.pdf.

Wayne, Lynda. 2005. "Metadata in Action: Expanding the Utility of Geospatial Metadata." Proceedings of GIS Planet 2005, International Conference and Exhibition on Geographic Information, Estoril, Portugal, May 30–June 2. http://cnre.vt.edu/gep/pdfFiles/Metadata_PDF%27s/2.2MetadataInAction.pdf.

6

INTERNATIONAL SPATIAL DATA INFRASTRUCTURES

Introduction

In chapter 5, we looked at national and state data portals and examined some associated issues such as metadata, standards, and spatial data infrastructure initiatives. In this chapter, we will look beyond state and national boundaries to the international context for sharing data and working with public domain spatial data.

Most of today's economic, health, climate, and environmental issues are global in scope, spanning administrative boundaries and transcending national sensitivities. Pollution, pandemics, rising sea levels, and famine are examples of issues that are *sans frontieres*—without borders. To help solve these global problems, decision makers need access to global data that are accurate, current, accessible, and objectively and multi-laterally describe the world we live in.

If the problem to be addressed crosses national boundaries, not only does the spatial extent of the data need to cross the same national boundaries, but the accompanying attribute data need to be defined in a consistent manner; the data should be collected under common rules with common sources and scales. Without a consistent approach to data capture and access to appropriate metadata, end users will spend valuable time and resources on projects simply trying to get the data to match, wasting time that should be spent analyzing data and making decisions.

Natural and human-made disasters and their aftermaths, as well as an increased awareness worldwide of the need for spatial data and GIS, have led to a number of international initiatives. These have already begun to help GIS practitioners effectively use public domain spatial data to address global issues.

Global spatial data infrastructure initiatives

As individuals and organizations began to realize their concerns were echoed by others around the world, the national spatial data infrastructure initiatives of the 1990s quickly expanded internationally in the 2000s and into the present decade. Remember from the last chapter and the discussion on National Spatial Data Infrastructure (NSDI) that such infrastructure involves more than just data. Actually, it may not contain the data at all, but provide an environment for spatial data to be more easily produced, shared, and used to their full potential. Indeed, a spatial data infrastructure is a "coordinated series of agreements on technology standards, institutional arrangements, and policies that enable the discovery and use of geospatial information by users" (Kuhn 2003). One of the best-known spatial data infrastructure initiatives is fittingly called the Global Spatial Data Infrastructure, or GSDI.

The GSDI

Decision makers and policymakers began to recognize during the 1990s that common standards for defining, collecting, maintaining digital spatial data, and promoting interoperable systems would encourage the effective use of GIS and spatial data. Around the time integration and sharing initiatives emerged in individual countries, people recognized that the same issues needed to be tackled on an international level, as well. Perhaps because the challenges encountered between agencies in the same country were multiplied a hundredfold when working across national boundaries, efforts moved slowly at first. Language differences turned out to be a relatively small challenge compared to the differences in pricing policies, security and access policies, national mapping and scientific organizational structures, definitions for mapped layers, standards and scales for collection, software used, and even ways of looking at the world geographically. However, the increasingly interconnected nature of global problems and the heightened awareness and urgency in addressing those problems, coupled with improved tools for sharing data, paved the way for a truly global spatial data infrastructure.

To create such an infrastructure required a great many meetings of minds. Indeed, from 1996 to 2002, six GSDI conferences were held, starting as a small group of interested parties and building to several hundred participants from more than fifty nations (Stevens et al. 2005). By 2005, nearly 1,000 attendees met at the eighth international forum in Cairo. The GSDI, like national initiatives, is concerned about the data along with policies, organizational structure, technologies, standards, financial resources, human resources, and delivery mechanisms that ensure data can be accessed and used. In short, the GSDI is concerned with everything that could eventually affect the use of spatial data for decision making. Somewhat paradoxically, the work of the GSDI isn't just about data but, on the other hand, it *is* all about the data, eventually. The goal is to enable data users anywhere in the world to access spatial data, at a variety of scales, and from multiple sources that ultimately will appear seamless to them. For this to happen, services, systems, software, and products need to be interoperable. Achieving this goal will depend wholly on enduring partnerships among public, private, academic, and nongovernmental organizations. This is a tall order, but the complex and pervasive issues we face as a global society provide the momentum required to drive it forward.

To keep GSDI goals moving forward, it is essential to persuade government officials that it is advantageous for them, from a cost and efficiency perspective, to not only invest in an SDI but also integrate their national efforts with the GSDI. Such an investment will reflect positively on the way their government is perceived, but it will also lead to better informed and more efficient decision making with a good return on investment (ROI). This may be most challenging in some developing countries where the transition from paper to digital maps is ongoing and where the work may progress in a rapid but uncoordinated and undocumented manner without the adoption of consistent standards. This echoes the experiences of more developed countries in the early decades of GIS. Involvement in GSDI without standards will not only inhibit sharing of data inside the country, but will also make data sharing even more difficult for international users. Persistent concerns about the security of information also remain. Recent reports about terrorist groups from one country using spatial data to plan and execute attacks in another country do little to allay these fears.

Despite progress, "most countries still lack a national framework to ensure that geographic information is consistent, available, and affordable" (Stevens et al. 2005). Establishing the appropriate partnerships and building those national frameworks is the only way forward to leverage resources and make the most of the data that are available.

The GSDI Association

The Global Spatial Data Infrastructure Association is a global consortium of organizations, agencies, companies, and individuals. When the association was incorporated in 2002, it gave GSDI initiatives a boost because, as an association, it could hire staff, build programs, and offer grants. The organization promotes efforts to hasten cooperation and collaboration and the ethical use of and access to geographic information. There are approximately forty full members, including the Chinese Taipei GI Center, the Department of Science and Technology in India, GISPOL in Poland (National Land Information System Users Association), the National Land Survey in Finland, HUNAGI in Hungary (Hungarian Association for Geo-Information), and Esri.

Due to the long-term nature of its goals, the GSDI Association works through permanent committees to carry out its mission, such as the UN Economic Commission for Africa (UN ECA), the Permanent Committee on GIS Infrastructure for Asia and the Pacific (PCGIAP), the Pan American Institute of Geography and History (PAIGH), the Economic Commission for Africa's Committee on Development Information Science and Technology (CODIST-II), the Open Geospatial Consortium, the European Umbrella Organisation for Geographic Information (EUROGI), and UNIGIS International.

ACCESSING THE DATA

The GSDI does not recommend any specific global data portal but does provide some suggested international, national, regional, and local data portals and repositories as a starting point for those seeking to access global geographic information. It also provides an area for organizations and countries to register their particular data portals and a library of literature related to global spatial data infrastructures.

INSPIRE

Recent devastating floods in many parts of Europe and the associated economic impact of past and potential flood damage, combined with increasing concerns about climate change, sustainable development, and biodiversity, made it clear to many in Europe that a European SDI was both necessary and long overdue.

The proposal for a directive establishing an Infrastructure for Spatial Information in Europe (INSPIRE), was first published by the European Commission on July 23, 2004, and came into force in 2007. The Commission's INSPIRE directive aims to make interoperable spatial information readily available in support of both national and European Union (EU) policy and to enable the public to access this information. INSPIRE also aims to solve data gaps, missing documentation, incompatible spatial datasets and services, and barriers to sharing and reusing spatial data.

INSPIRE will be implemented in five stages, with full implementation required by 2019. It will establish an infrastructure for the twenty-seven member states to standardize the collection, collation, and sharing of spatial data by public sector organizations, to support regional cross-boundary environmental policies or activities that have an impact on the environment. These are the five common principles of the directive:

1. Data should be collected only once and kept where it [data] can be maintained most effectively.
2. It should be possible to combine seamless spatial information from different sources across Europe and share it with many users and applications.
3. It should be possible for information collected at one level or scale to be shared with all levels or scales; [These data should be] detailed [enough] for thorough investigations, [while being] general [enough] for strategic purposes.
4. Geographic information needed for good governance at all levels should be readily and transparently available.
5. [It should be] easy to find what geographic information is available, how it can be used to meet a particular need, and under which conditions it can be acquired and used.

   Source: European Commission INSPIRE website.

Essentially, INSPIRE is the glue that binds all the environmental geodata from each of the EU's national mapping agencies and public authorities. As the directive is adopted by the participating national legislatures, every house, lamppost, phone mast, roundabout, river, and host of other phenomena will be connected with data on transport networks, place names, postcodes, population statistics, and environmental indicators. It represents one of, if not the most far-reaching spatial data infrastructure initiatives in Europe.

The directive also requires the EU to establish a community geoportal through which member states would provide access to their data and other portals they operate. A vital component of the geoportal would be the collection and maintenance of metadata; the metadata regulation documents the requirements for the INSPIRE directive that go beyond, but comply with, many existing international metadata standards. Some of these data are already online on the European Environment Agency's Data and Maps portal with such themes as agriculture, air pollution, industry, soil, waste and material resources, and household consumption.

With the infrastructure came eSDI-NET+, a framework for cross-border dialogue and exchange of best practices on SDIs throughout Europe. Its goal is to bring together key SDI participants and stakeholders, promote high-level decisions and bottom-up technical discussions, help geographic information (GI) stakeholders realize the full potential of digital GI for content providers and users, and increase awareness concerning the importance of GI. It focuses on maximizing benefits of a number of spatial data initiatives and e-government programs, and seeks solutions for multicultural and multilingual access, exploitation, use, and reuse of digital GI content in Europe. The most important goal of eSDI-NET+ is probably to aggregate existing national datasets of core GI into cross-border datasets.

ACCESSING THE DATA

The spatial data under the remit of the directive is extensive and varied, categorized into thirty-four themes, from protected sites, habitats, and biotopes to area management, population, and health and safety. Where possible, the INSPIRE geoportal website will link to existing national portals, and will not store or maintain the actual data. At the time of writing, the current version of the portal is a prototype for discovering and viewing datasets and services.

The role of international standards organizations

As we have seen, GIS has evolved from individual systems in individual organizations to a global phenomenon. The requirements to share models, programs, and data across departmental, organizational, and ultimately international boundaries, have also evolved. As a result, international standards organizations are more and more important to the advancement of GIS. These organizations have an impact on anyone using public domain spatial data, as they are responsible for developing and implementing guidelines for working with the data. While they do not, or cannot, dictate the regulation surrounding data use, they nevertheless influence how data should be made available, what attributes should describe data themes, and much more.

One such organization is the International Organization for Standardization (ISO). ISO is the world's largest developer and publisher of international standards and its scope goes way beyond GIS alone. It works with a network of the national standards institutes of 163 countries, but is, itself, a nongovernmental organization. ISO has a central secretariat in Geneva, Switzerland, that works to meet the needs of both public and private interests. ISO has standards related to GIS, including ISO standard 19111 of 2007. It defines the "conceptual schema for the description of spatial referencing by coordinates, optionally extended to spatio-temporal referencing. It describes the minimum data required to define one-, two-, and three-dimensional spatial coordinate reference systems with an extension to merged spatial-temporal reference systems. It allows additional descriptive information to be provided. It also describes the information required to change coordinates from one coordinate reference system to another." Anyone who has worked with spatial data knows how important it is to convert between different coordinate systems; this is just one example of how ISO standards have a direct impact on GIS.

Another key organization is the Open Geospatial Consortium, Inc. (OGC), a nonprofit, international, voluntary consensus standards organization that leads the development of standards for geospatial and location-based services. Unlike ISO, OGC is entirely focused on GIS and geospatial technologies. Started in 1994, it now includes more than 400 commercial, governmental, nonprofit, and research organizations around the world. Most of the OGC standards depend on a generalized architecture captured in a set of documents collectively called the "Abstract Specification," which describes basic data models for representing geographic features. On top of the

Abstract Specification, OGC members are developing a growing number of specifications or standards to meet specific requirements for interoperable location and geospatial technologies.

Due to the dramatic expansion in GIS services, tools, and data available in many formats on the web, another standards organization that influences GIS is the World Wide Web Consortium (W3C). Its mission is to "lead the World Wide Web to its full potential by developing protocols and guidelines that ensure the long-term growth of the Web." XML technologies, accessibility for those with disabilities, semantics, and web design and architecture are a few of the efforts W3C is involved in. Of particular impact on GIS is XML, since it has been adopted as the primary technology for creating, storing, and transferring spatial metadata. Other standards organizations that are important to GIS are the International Hydrographic Organization, the Web Services Interoperability Organization, the Location Interoperability Forum, the Wireless Location Industry Association, and, as we discussed earlier, the Global SDI.

Finally, some concerns about standards have existed since the inception of GIS and indeed, transcend GIS to include any technology or methodology that involves the international community. The adoption of any standards requires a tradeoff between functionality and performance. Standards aim to make operations more efficient and provide for broad functionality, and acceptance, of goods and services. For example, adopting standards for spatial data interchange formats enabled access to a broad range of data sources across operating systems that were themselves based on standards. Having standard vector and raster formats, even though they are numerous, has always allowed GIS users to share data. However, standards are, by their very nature, inflexible. They do not allow fine tuning to specific software, hardware, or applications, and particularly nowadays, cannot match the rapid rate of technological change and innovation. Some de facto standards in GIS are not necessarily efficient, nor the best available and certainly no longer state-of-the-art, but may continue to exist simply due to the original popularity of a particular method, hardware, or software.

Global spatial infrastructure initiatives: the issues

Previously, the lack of a formal standard or any central policy on the collection, storage, and access to geographical data inhibited the uptake and wider adoption of GIS. So why have a set of proposals, based on principles that seem largely self-evident, proven to be fairly contentious and taken so long to get to this stage?

Bregt and colleagues (2009) note that establishing an SDI that crosses national boundaries is a complex endeavor fraught with difficulties. Among them are standardization and data handling intertwined with assumptions and beliefs with which the data were collected and stored. Crompvoets and colleagues (2004) stated that there was a declining trend in the use, management, and content of national clearinghouses, one of the main elements of SDIs. Craglia and colleagues (2008) observed that in spite of a greater emphasis on interoperability through services, the underlying basic approach to an SDI architecture has not evolved much during the last decade. Is there some element of wishful thinking in the claims of many countries, along with some overly optimistic rhetoric of being involved in some form of SDI development (Masser 2005)?

For example, INSPIRE is based on the belief that good quality, harmonized geographic information is needed for good governance across Europe. The principles of INSPIRE are that geographical information should be collected once and maintained at the appropriate level. It should also be seamless, shared, available on terms that do not constrain its use, easy to find and assess, and presented in an understandable visual form—all laudable aspirations. However, one of the most contentious problems has been the terms under which the public can access the data.

Under current proposals, spatial data collected by the public sector organizations and national mapping agencies would remain under the ownership of those organizations and agencies and would not be placed in the public domain. As such, anyone wishing to download data discovered on the INSPIRE geoportal would have to contact the source organization or agency to arrange access to and, if necessary, pay for the data and comply with copyright restrictions. Many believe that data created by the public sector should be usable by the general public at no cost, or minimal cost. They regard the proposals as perpetuating the outdated policy of charging the citizens of Europe for information they have already paid to collect.

Some private-sector firms that work in this area object on grounds they could not afford the government-sponsored competition they would face. Some street mapping firms have been particularly active in their lobbying against the data access proposals. In addition, there is the question of what data should be collected. During the consultation period for INSPIRE and other initiatives, stakeholders are asked for the type of data they require. It is not feasible to collect all the data that will be requested by every respondent; ultimately, a board of directors must decide what data should be collected and by whom. Some of the respondents may have the power and influence to see that their interests are safeguarded; others will have to make concessions and sacrifices. Over time, INSPIRE has changed its emphasis on licensing and copyright, even for viewing of data, and some of the essential datasets that it was meant to cover have been removed from the list.

Defra (Department for Environment Food and Rural Affairs) is the lead government organization for implementing INSPIRE in the United Kingdom. In a paper published in 2003 (INSPIRE 2003), it lists five obstacles to accessing and sharing data:

1. Spatial data gaps
2. Inadequate documentation
3. Incompatible spatial datasets
4. Incompatible geographic information systems
5. Sharing and reuse barriers

Others point out that the development of open infrastructures for spatial information will require sustainable funding by public authorities at the local, national, and European levels, especially during the initial phases. While it will, in the future, save money at many levels of administration due to the streamlining of tasks and sharing of information, these benefits are difficult to sell legislatures and parliaments that may only look at short-term gains. The NSDI in the United States languished for that very reason; the FGDC was given a very small budget and borrowed staff from existing agencies.

Advocates for SDIs argue that their interests should be treated the same as physical infrastructures that are built, maintained, and operated by the public sector. Spatial data infrastructures, they say, need to be developed on top of the existing SDI foundations in member countries and they must be funded and supported for the long term, as they are just as vital to the economic well being of society as roads, bridges, and buildings. If public funds are used to build infrastructures for spatial information, then they must be available for the benefit of and use by all of society. Conversely, the producers of public spatial data are deprived of any revenue generating opportunities and must be compensated.

Key implementation issues related to establishing organizational structures need to be addressed:

How could the private sector be involved? Should the private sector be involved?

What should the roles be for research and development and education and raising awareness?

What should be the priorities for implementation, and what should be the underlying technological infrastructures?

What software should be used?

How should multiple languages be accommodated?
What common codes for attributes should be used?

It is also important to clarify in the INSPIRE legislation how the infrastructure for spatial information, which will initially focus on information needs for environmental policies, will expand to meet the requirements of other sectors over time. Also, the ongoing relationship between INSPIRE, governmental, and IDA (Interchange of Data between Administrations) initiatives has been the focus of recent attention. Each requires clear and unambiguous remits to avoid duplication and over-management.

The emphasis of the INSPIRE's directive has moved towards a policy based on intellectual property rights held by government agencies over the publicly funded data they collect. The directive calls for affordable data and transparent pricing. For potential data users, this and other issues should be of concern. First, what, if any, layers or components of the data produced under INSPIRE will be public domain? Second, will there be any charges for public use of these data? If so, how much should be charged and under what terms and conditions?

In January 2006, the Council of the European Union formally adopted a common position on the INSPIRE directive, stipulating that geographic data collected by national mapping agencies all over Europe should be owned by such agencies and not be in the public domain. As such, restricted access to geographic data for both the public and commercial enterprises, due to high costs and restrictive licenses, would ultimately mean fewer services and opportunities in Europe. Many believe if the European Parliament does not adequately amend or, failing that, reject this stipulation, INSPIRE will entrench a policy of charging citizens for information they have already paid to collect. At present, the rules governing access to the data remain undefined, although all requests for public access to the data are to be considered on a case-by-case basis.

FOR FURTHER READING

To learn how INSPIRE developed and how it would be implemented, look at *Building European Spatial Data Infrastructures* (Esri Press 2010) by Ian Masser. To find out more about the concerns surrounding INSPIRE, see the Open Rights Group discussion on INSPIRE.

International data portals

An increasing number of international spatial data portals have emerged to provide information on a variety of themes, including climate, hydrology, geology, soils, energy, and population. The following section will review a selection of the more prominent and enduring data portals and discuss some of the issues that led to the creation of the portals in the first place.

Some international data portals predate international data-sharing infrastructures or are only peripherally related to these initiatives; in some cases, the infrastructure has widened to include some of these portals. In any case, the portals are just as useful as the initiatives to anyone trying to access global spatial data, and in some cases even more so.

Global Geographic Information System

During the 1990s, the USGS and the American Geosciences Institute (AGI, formerly the American Geological Institute) recognized the restrictions facing many organizations that sought to access and work with geographic information: costly hardware and software platforms, expensive data, and a lack of technical expertise. The USGS and AGI also identified the need for a more integrated approach in data provision and the extra value that could be gained by combining stand-alone datasets into integrated data repository.

To this end, the USGS and AGI announced in 2000 the release of Global Geographic Information System, a 1:1 million-scale database of land cover, elevation, and natural resources. The database, reproduced as a complete collection or by region (on CD and DVD) combines and enhances the value of individual datasets and comprises twenty-eight vector and eight raster datasets.

Vector data include
- political boundaries and major cities
- transportation infrastructure—roads and railways
- geologic age, perennial rivers, and drainage basins
- ore deposits, climate data, and volcanoes

Raster data include
- shaded relief
- population

- slope
- land cover

This dataset is unique as it comes from public domain sources, is global, and contains data layers that are not easily found anywhere else, such as pipelines, mineral and ore deposits, and surficial geology.

ACCESSING THE DATA

Data are available from the Global Geographic Information System website as shapefiles, rasters, and georeferenced images, and come packaged with a customized version of ArcReader so non-GIS users may examine the data.

PIGWAD

Undoubtedly the largest portal in terms of its spatial extent, the Planetary Interactive GIS on the Web Analyzable Database (PIGWAD) was launched by the USGS in 2000. The aims of this project include helping the planetary research community apply GIS and geospatial open standards and creating planetary spatial databases that include geological features and topographic maps.

A variety of vector and raster data for a number of planets are available to download. For example, the H-1199 Geologic Map of the Tolstoj Quadrangle of Mercury, the G-2534 Geologic Map of the Galileo Regio Quadrangle of Ganymede (a moon of Jupiter), the topographic orthophotos of the Moon, and the DTMs for Mars.

To promote the use of the PIGWAD data and GIS, USGS also provides a number of tutorials and a supporting website.

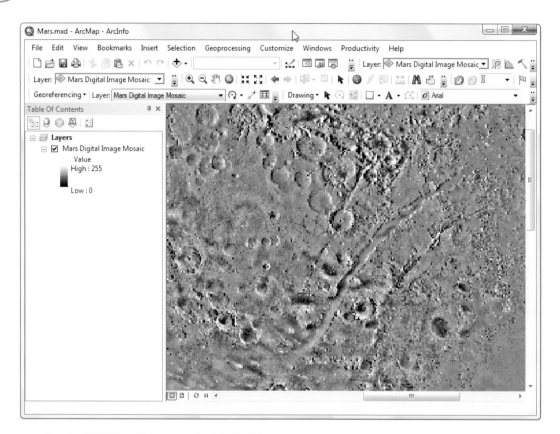

Sample of USGS's Mars Digital Image Mosaic (MDIM2.1). Viking Mars Digital Image Mosaic (MDIM2.1) courtesy of NASA/USGS Astrogeology Science Center.

ACCESSING THE DATA

The vector data are available to download from the PIGWAD website as zipped shapefiles with accompanying metadata. Raster data are available to download as orthorectified JPEG images, GeoTIFFs, and TIFFs.

CIESIN World Data Center

The Center for International Earth Science Information Network (CIESIN) World Data Center is one of the earliest international data portals and dates back to the early 1990s. Hosted by the Earth Institute at Columbia University, it is one of fifty-one such data centers within the World Data Center System. The CIESIN data center provides access to data relating to human interactions with the environment, such as flood hazards, landslide risk assessments, population, biodiversity, and agriculture. The site hosts services for searching and downloading data, and provides online tools for visualizing, converting, and querying the data. The data are structured into thematic portals, including climate, health, and population to facilitate searching. Each thematic portal includes an option to download datasets, both tabular and spatial. As with many data portals, the site lacks a task-oriented approach to locating and downloading the large volume of available data.

ACCESSING THE DATA

Spatial data are available to download from the CIESIN World Data Center website in shapefile, GRID, and JPEG format.

United Nations FAO GeoNetwork

The Food and Agriculture Organization (FAO) of the United Nations was responsible for the development of the GeoNetwork data portal. GeoNetwork aims to improve access to, and integrated use of, spatial data while promoting a multidisciplinary approach to sustainable development. Using a combination of spatial and attribute searches, visitors to the site can search for and preview (where available) both the data and the metadata. The site also includes useful links to social networking sites such as Facebook and Delicious for bookmarking datasets of interest.

These are the core datasets available via the GeoNetwork portal:

- global administrative boundaries
- land cover and land use
- soils and soil resources
- forestry and aquaculture
- climate
- population and socio-economic indicators
- biology and ecology

The GeoNetwork site also provides access to supplementary publications and technical reports.

ACCESSING THE DATA

Much of the spatial data on the GeoNetwork website are available as online low- or high-resolution images or available to download in choice of formats—grids, shapefiles, and a number of raster formats.

UNEP Geo Data Portal

The UNEP Geo Data Portal was established in 2000 to "provide a comprehensive, reliable and timely supply of data" used by both the UNEP and its partner organizations in the preparation of the UN's Global Environment Outlook (GEO) report. To support the GEO report, this diverse range of spatial data was required to be

- available with global coverage, but with data at the national level;
- uniform in both data definition and collection standards;
- available continually since 1970;
- free and accessible to the GEO partners and the wider community; and
- available as aggregated values for sub-regional, regional, and global levels.

This repository has grown to hold information on over 500 environmental themes such as population, emissions, disasters, health, and hydrology. The UNEP Geo Data Portal site only supports attribute searches at present; once you have located a dataset or datasets of interest, the data may be viewed online or downloaded as required.

ACCESSING THE DATA

The data hosted on the UNEP Geo Data Portal website can be viewed online or downloaded in a variety of formats including ArcInfo export files, TIFF and BIL image formats, ASCII grid, and Esri shapefile. Due to copyright restrictions, some of the data are not publicly available to download. Any data downloaded from the site and reused elsewhere must always be cited.

TerraViva! GeoServer

TerraViva! GeoServer is hosted by ISciences, LLC., a research organization that promotes a better understanding of earth processes and sustainable development and aims to provide "state-of-the-art information about the earth and its people" through the development of software tools to improve public access to data.

Spatial data are available under the following categories:

- anthrosphere
- atmosphere and climate
- biosphere
- geosphere
- geostats databases
- hydrosphere
- maps

Using the TerraViva! Interactive Map Viewer, it is possible to compile maps online and export the data in PDF format.

ACCESSING THE DATA

Data available from the GeoServer portal remain subject to national and international copyright laws and, as such, the source must always be cited. Although the data can only be downloaded in the ISciences proprietary map file format, XTVM, a free Global Data Viewer is also available to download. Lower resolution data are available to download for free; higher resolution data are available for purchase.

GEOSS GEO Data Portal

The Global Earth Observation System of Systems (GEOSS) represents a unique and ambitious project to link a range of earth observing systems including ocean buoys collecting temperature and salinity data, seismic monitoring stations, and satellites remotely sensing a range of environmental phenomena. The aim is to provide a global public data infrastructure that generates comprehensive, near-real-time environmental data, information, and analyses. In recognizing the need for a common standards data-sharing, each contributor to the project must sign up to the GEOSS data-sharing principles to ensure full and open access to data and metadata. Seventy-four countries and fifty-one participating organizations are currently involved in the project.

GEOSS identifies nine distinct groups and data users: disasters, health, energy, climate, weather, water, ecosystems, agriculture, and biodiversity. To support the work of those users, the information is made available via the GEOSS Geo Portal, a data clearinghouse powered by Esri Geoportal Server. This provides the framework to search and identify a variety of datasets through an ArcGIS-server-powered web GIS interface.

ACCESSING THE DATA

Data are not available to download directly from the GEOSS website; its primary function is to allow users to connect to databases, services, and other data portals to access the most up-to-date and reliable earth observation data.

Europe has developed an equivalent earth observation system called the Global Monitoring for Environment and Security (GMES) program. GMES collates data from a variety of earth observation sources, such as airborne and seaborne sensors and ground monitoring stations, to provide a thematic project-based framework for locating relevant environmental information. The six thematic areas are marine, land, atmosphere, emergency, security, and climate change.

Global Water System Project

The Global Water System Project (GWSP) has developed a digital water atlas describing the basic elements, interlinkages, and changes by creating a consistent set of annotated maps that transcend national boundaries. The data are in-kind contributions from the partners and are hosted by the University of Bonn, Germany. Data include water consumption of power plants, annual water withdrawals for households and commercial use, annual river discharge, and a climate moisture index.

ACCESSING THE DATA

To access the datasets hosted on the GWSP website, the user is directed to contact the individual or organization responsible for those data.

Working groups versus data portals

It is important to distinguish between international working groups and resources from which spatial data may be obtained. Most of the working groups and think tanks develop policy, standards, and collaboration among agencies. Given the issues we have examined that could prove to be barriers to collaboration, such as copyright, fees, and common definition of mapped features, this behind-the-scenes work is critical to the success of making international spatial datasets available. However, few of these sites actually host spatial data, and if they do, it is via a web browser without the capability to download.

One example is the United Nations Geographic Information Working Group. It was formed to address geospatial issues and develop common geographic databases but does not host its own spatial data. However, both it and other international policy organizations provide a good starting point to search for other international spatial datasets; for example via the portals page (navigate to Data portals and links). Datasets referenced include administrative boundaries, names, disasters, and a remote sensing data archive. Some data referred to on the portals page are directly accessible, while others, for example, the international boundaries, are only available to United Nations staff. Although users who do not work for the United Nations are provided with a UN contact, access to the data is not guaranteed.

Another example of a working group site hosting spatial data is the World Health Organization's GeoNetwork. This site is so large that a careful selection of keywords to search on is critical. It does however, provide a wealth of data. UN organizations including the FAO, World Food Network, and Environment Programme use the same software to access the data from a mapping applet and from direct links, including digital maps, satellite images, and statistics, following the standards set by the International Standards Organization and the Open Geospatial Consortium. Much of the data, such as health facilities, are available but potential data users are required to contact the dataset owner. Some of the data are copyrighted.

Data portals based on common issue or region

Some international data portals evolved from a common theme, often an environmental issue (frequently an issue that crosses national boundaries) or an entity, region, ecoregion, or a body of water. In the following sections, we will examine a number of these portals.

Atlas of the Biosphere

The Atlas of the Biosphere, hosted by the University of Wisconsin–Madison's Center for Sustainability and the Global Environment (a part of the Nelson Institute for Environmental Studies), includes data on human impact, land use, ecosystems, and water resources. It was created from a variety of sources, including the Population Reference Bureau, the University of Kassel (Germany) Center of Environmental Systems Research, NASA, the University of East Anglia's Climate Research Unit, and other government and university sources.

ICES

The International Council for the Exploration of the Sea (ICES) "coordinates and promotes marine research on oceanography, the marine environment, the marine ecosystem, and on living marine resources." The ICES is a network of 1,600 scientists and 200 institutions linked by an intergovernmental agreement to gather information and to conduct research. Members include all coastal countries bordering the North Atlantic Ocean and the Baltic Sea, with affiliate members in the Mediterranean Sea and southern hemisphere. Its data holdings include ocean measurements of conductivity, temperature, weather, and other variables on the surface and under the surface from sensors, instrument bottles, and ships.

HELCOM

The Helsinki Commission (HELCOM) is the governing body of a convention organized in 1974 to protect the marine environment of the Baltic Sea. GIS was a natural solution for HELCOM, as it must support decisions about complex issues such as marine accident rescue and environmental quality, and do so with the cooperation of the ten

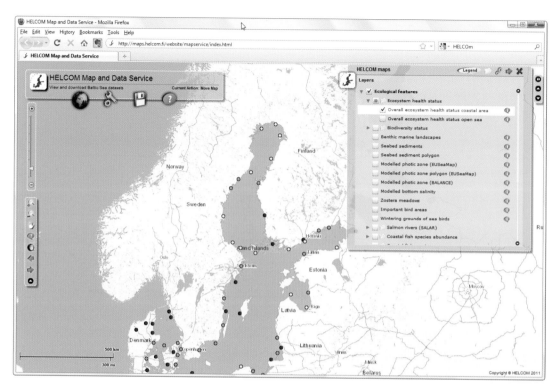

HELCOM's Map Service. Copyright HELCOM 2011.

countries that border the Baltic Sea. HELCOM uses GIS to support interactive web-based maps such as the Marine Accident Response Information System, Sea Track Web (for oil spills and oil weathering), the marine spatial planning tool, the Baltic nutrient GIS for discharges, the cooperative monitoring program, the coastal fish monitoring GIS, and the atlas on protected areas.

HELCOM also maintains a GIS data delivery service, which allows users to view and download dozens of layers, including ecologically important areas, shipping and oil, monitoring, pollution load, background data such as borders, territorial boundaries, and depth contours, and shipping traffic estimates. The service was initially released on an ArcIMS platform but has been rereleased in 2010, and greatly enhanced, on an ArcGIS for Server platform.

ACCESSING THE DATA

Vector data are available to download from the HELCOM map data delivery service website in shapefile format and raster data are available in TIFF format.

Arctic Portals

The Arctic Council was established in 1996 as "a high-level intergovernmental forum for promoting cooperation, coordination, and interaction among the Arctic States." It deals with common Arctic issues, such as sustainable development and environmental protection in the Arctic and the involvement of the Arctic Indigenous communities and other Arctic inhabitants. The current focus on climate change and the desire to quantify the as yet untapped natural resources in the Arctic have generated a wide array of spatial data. The collection and management of those data had mostly been undertaken at the national level and, as a result, datasets were distributed across many organizations in an unintegrated and poorly coordinated manner (Barry 2008). The Arctic Council, and others, recognized the need for a dedicated Arctic spatial data infrastructure to provide the necessary framework to support better integration of and access to those datasets.

In support of this initiative, the Arctic Council developed the Arctic Portal Mapping System to provide visual geographical information about Arctic-related data, such as distribution of vegetation, permafrost, spread of indigenous languages, and information about Arctic affiliations and organizations. Data are provided through various affiliations and organizations working in collaboration with the Arctic Portal, such as the University of the Arctic, The Circumpolar Arctic Geobotanical Atlas, International Permafrost Association (IPA), Conservation of Arctic Flora and Fauna (CAFF), Protection of the Arctic Marine Environment (PAME), and UNEPGRID-Arendal.

The Arctic Portal Interactive Web Map allows for upwards of thirty map layers to be examined. Although the map created online may be exported as a graphic (JPEG format), at the time of writing, no spatial data may be downloaded.

In a parallel effort, the USGS has funded a research project, the Virtual Arctic Spatial Data Infrastructure, to collate the available Web Map Services (WMS) in the Arctic region and to prototype a viewer application to relay multiple WMS from their original host sites. The primary aims of the project are to demonstrate availability and gaps in Arctic map services, register existing WMS instances in the GEO/GEOSS global service registry, and promote a wider understanding of WMS standards and techniques.

ArcticWeb, a related project developed in Norway, provides information for the energy industry on exploration and production licenses, petroleum activity, ship movement and emergency equipment. The contents of ArcticWeb are sourced directly from public databases and presented in a format designed for the energy industry. Given its focus, it is understandably sponsored by six oil companies: Statoil, ConocoPhillips, BG Group, ENI, Lundin, and Shell. However, interest in the data goes beyond the energy industry. The goal of the project is to create a web framework that would continuously harvest predefined information from data owners—reputable authorities and institutions that produce and maintain reliable and high-quality information related to Arctic areas. These owners include the Norwegian Coastal Administration (Kystverket), the Geological Survey of Norway (Norges Geologiske Undersøkelse), and other government agencies.

International sea level data

The British Oceanographic Data Centre (BODC) provides access to marine environment data and aims to promote a better understanding of the processes that control the natural environment by helping to answer both local (coastal flooding) and global (climate change) questions. Some of the global data include the following:

- Argo floats (a collection of small, drifting probes deployed worldwide that collect temperature and salinity profiles from the top 200 meters of ice-free global oceans)
- bathymetry data
- international sea level data
- conductivity, temperature, and depth profiles for the Atlantic Meridional Transect (an annual oceanographic research project from the United Kingdom to the Falkland islands)

Access to some of the data is restricted to users registered with the site and remains subject to tighter licensing controls as to the further use of the data.

ACCESSING THE DATA

Data are available to download from the BODC website, in Grid, NetCDF (Network Common Data Form), or ASCII format, or via an online request facility.

The North American Land Cover Database

The North American Land Cover Database is another example of an international initiative developed by organizations with a common requirement for spatial data on a theme that transcends national borders—in this case, land cover. Natural Resources Canada, the Canada Centre for Remote Sensing (NRCan/CCRS), the USGS, and three organizations from Mexico—the National Institute of Statistics and Geography (Instituto Nacional de Estadística y Geografía, INEGI), the National Commission for the Knowledge and Use of Biodiversity (Comisión Nacional para el Conocimiento y Uso de la Biodiversidad, CONABIO), and the National Forestry Commission of Mexico (Comisión Nacional Forestal, CONAFOR)—established the North American Land Change Monitoring System (NALCMS). The project is facilitated by the Commission for Environmental Cooperation, an international organization created under the North American Agreement on Environmental Cooperation to address regional environmental concerns, help prevent potential trade and environmental conflicts, and promote the effective enforcement of environmental law.

The common goals of this multi-scale, land cover monitoring framework—applied across North America and meeting the requirements of each individual country—have been partially met through the first product of the collaboration, the 2005 Land Cover Database of North America. This dataset, generated at a spatial resolution of 250 meters, is based on observations acquired by the MODIS satellite. The mapping was performed individually by the above organizations but the classification scheme (designed in three levels and based on the FAO Land Classification System) united their efforts. Levels I and II are common throughout North America while level III is country-specific. The resulting data can be used to assess climate change, carbon sequestration, biodiversity loss, and changes in ecosystem structure and function.

ACCESSING THE DATA

The North America Land Cover Database can be accessed through portals maintained by each agency involved in the initiative.

SDI progress in Africa

In Africa, SDI is regarded with anticipation by many who see it as almost a prerequisite for the development of the continent. The scarcity of financial resources focuses attention on any approaches that maximize investment, with spatial data being collected once and used in as many applications as possible. According to Ezigbalike (Verduyn 2009), every delegate to the Committee on Development Information, Science and Technology (CODIST-Geo)—a body that reviews challenges and issues relating to information and communication technology (ICT), geo-information, science and technology to help address Africa's development challenges—agrees that the way forward in providing geo-information services to their constituents is through the SDI framework. However, he states this appreciation of an SDI's benefits is still largely confined to members of core geo-information industries and has yet to permeate allied disciplines. Funding also remains an issue.

Nevertheless, progress has been made through the work of several intergovernmental organizations. One is the Regional Centre for Mapping of Resources for Development, which promotes the development and use of geo-information and IT in the sustainable development of Africa. Another is the Regional Centre for Training in Aerospace Surveys, a joint project among several countries providing training, research, consultancy, and advisory services in geoinformatics. Both organizations raise awareness and organize specific courses for practitioners.

Limitations of international SDIs and portals

Despite all of the progress that SDIs have made over the past decade, some people remain skeptical as to the value of their contribution. Koerten (2008) goes so far as to state that SDIs have actually hindered the implementation of portals: "The advancement of SDI has made implementation processes more complex. Probably due to the fact that SDIs are moving across organizational boundaries, appealing projects take longer than expected, are delayed or sometimes even cancelled. The most striking fact is that, explaining setbacks and failures, practitioners point at organizational impediments, but do not know how to deal with them." His point is not to discourage the development of SDIs, but simply to note that coordination across international boundaries and among organizations is inherently more difficult than working within single countries and national organizations.

What are the limitations of international spatial data portals? One common limitation has to do with how the user interacts with the portal through the web interface. A recurrent criticism of many data portals is that they are poorly designed and difficult to use. Some appear to have been set up by those who are technically and programmatically competent but, as they are not data users themselves, the portals tend to lack an easy-to-use interface and also lack the best, time-saving search tools for the job. Some portals are big on aesthetics but short on useful functionality; just because the latest web widget supports some "cool" graphics doesn't mean they are appropriate for a data portal site. Other portals are unreliable and lack any backup system, such as an FTP site where data can be downloaded if the main website is unavailable or bandwidth is restricted. Some are full of news about spatial data but actually contain very little spatial data themselves, while others have online web mapping capabilities but again, little in the way of data that can be downloaded and used by a GIS analyst. In the following sections, we will look at some of the main issues facing international spatial data portals.

Data duplication

Perhaps predictably, there does seem to be some duplication of effort, as in the case of the GEOSS and GMES portals. Is this a feature of the inherent suspicion and mistrust of some other organization's monitoring? Are these initiatives in danger of repeating the old data silo problems already experienced at the local and national levels? Individuals and organizations skeptical about the capabilities and focus of other organizations often decide to go it alone to ensure they have control over and ready access to the data and are not reliant on the priorities and schedules of others. Will ever-present economic, political, military, and national security threats always work to defeat multilateral and sustainable access to international data reserves? Will we truly have access to global data, or will that access be (and should it be) restricted to key decision makers?

Standards and interoperability

Individual countries have different concerns and priorities. They have to prioritize their data collecting and sharing initiatives according to national objectives, hopes and aspirations; these may be at odds with the objectives of their neighbors and other countries. International data-sharing initiatives will work *if* they have national and local support and *if* all participating countries and organizations agree to adopt the same standards. With so many data portals now available, hosting huge reserves of data for individuals and organizations to work with, the requirement for international standards has never been greater. Organizations are able to share, coordinate, and communicate with GIS in a common data infrastructure, but only if those datasets are interoperable. To attain this level of interoperability, an agreed set of standards becomes the language by which different departments and organizations communicate. In so doing, a more open GIS community is created, which allows for the sharing of spatial data, integration with different GIS technologies, and integration with non-GIS applications. It can operate on different computers, running different operating systems, on the web, or on mobile devices.

The need for standards was recognized in the early days of GIS, but the primary concern then was engaging with those using other types of GIS software. The ability to convert formats such as GIRAS, MOSS, DLG, and other commonly used files into something that other GIS software could use, was paramount. Sharing spatial data with other core business applications, even within the same department, was fraught with difficulty and as a result, GIS remained rather disconnected from other functions of most organizations. Over time, the GIS community recognized the need for standard interchange formats (such as SDTS, GML, and DXF), open file formats (such as shapefiles), direct read application programming interfaces (APIs), common features in a database management system (such as the OGC simple feature specification for

SQL), and integration of standardized GIS web services, including ArcIMS, WMS, and WFS. With this evolution, GIS has become embedded into standard business practices and the wider information technology infrastructure.

Perhaps unknowingly, these developments have a direct impact on anyone using public domain spatial data today. Without adherence to standards and open GIS, users would be faced with the same issues now as those working in GIS in the late 1980s—when there was much less sharing of spatial data and an uncoordinated and unsustainable approach to decision making.

Scope of international data portals

We have chosen to describe certain data portals in this chapter because they illustrate specific trends in public domain spatial data provision in terms of data access, licensing, themes, and regions covered. Some, like ArcticWeb, began as a partners-only, restricted access library, with partners contributing time and financial resources to the project. In 2010, ArcticWeb removed its access limitations so that more users may now have unrestricted access to its resources, as guests and as registered users. An increasing number of portals maintain two sets of resources, basic and premium. Basic resources are offered either for users without an account or those logging into a general guest account, and are generally available for free. Premium resources are typically offered on a fee basis for those who register with the site, and on some portals, where the registration is only open to approved stakeholders. For those portals with two different levels and a fee-based service, it is often the classic case of "you get what you pay for." As you examined in chapter 4, it is sometimes worth paying a fee and saving yourself, or your organization, hours or days of processing to get exactly the type, format, and scale of data that you need.

A number of other portals are now providing their premium resources for free, with general access provided for all levels of users. One important limitation that remains in place for most data portals, and a key consideration for end users, is the restriction on using the data for commercial and noncommercial purposes.

Once placed online, data portals often expand in terms of the number of contributing partners involved, the numbers of data layers available, and the geographic extent they cover. However, once a portal is set up, other interested parties often come forward from within the original target region and from outside that region. For example, although at the present time ArcticWeb covers only the Norwegian Continental Shelf, discussions about including other areas in Scandinavia are currently underway. The geoportal provides access to data from many fields, including biology, geology, meteorology, and oceanography. It is quite likely that other stakeholders will advocate for their own data to be included. The managers of ArcticWeb will then face the same dilemma other portal creators have faced—does all collaboration benefit a portal project, or is some collaboration too much? Is there a point where more collaborators dilute the original intent of a particular portal to the extent that it becomes unmanageable? Will additional collaboration and the requirement to host new datasets jeopardize the needs of the original collaborators or make the original collaborators wary about committing resources to it in the first place?

Instead of trying to be all things to all people, smaller, more focused international data portals designed with the end-user in mind and targeted at specific, common issues which affect us all (climate change, environmental, economic, transport, and so on) seem more likely to succeed. The HELCOM and Arctic Portal Mapping System are good examples of international data portals with very specific remits and objectives.

That said, even some narrowly focused data portals have failed to last the distance. In chapter 5, we noted the demise of the Conservation GeoPortal. Among the reasons given for the withdrawal of the site was insufficient funding to maintain and update the resources. In a rather telling summary of their experience, the organization noted that although the GeoPortal was visited frequently by people to search for data, very few people posted their own metadata. Their reflection was that the concept of sharing data is much more advanced than the practice. Even the Atlas of the Biosphere, mentioned earlier in this chapter, for all of its richness, was originally set up by a student at the University of Wisconsin–Madison as part of a master's degree project. Who will update this database in the future? Will the university be interested in maintaining the server where it is hosted? Does the university know that this dataset is hosted on its server?

During 2009, it was announced that after five years, the data service Platial had been retired and the data donated to Geocommons, where it is now available under a Creative Commons license. Some of its functionality went to Google's My Maps. Users were advised to move their data from Platial's servers as soon as possible. For those with Platial maps on another site, there was no solution for easy migration.

Ensuring appropriate use

We have mentioned many times in the previous chapters how important it is for the data to be fit for the user's intended purpose and that is it the user's responsibility to ensure the data are suitable for the intended application. This becomes even more of a challenge when the end-user is far removed from the source of the data. If issues arise and there are no local contacts, it can be a frustrating experience dealing with differences in time zones, differences in interpretation when translating the supporting documentation, or different expectations as to what is available.

Although such events are rare, the inappropriate use of spatial data may have grave consequences. In 1998, a US Air Force aircraft on a training exercise in Northern Italy severed the cable of a ski resort gondola, causing the gondola to crash to the ground, killing all twenty people on board. During the subsequent investigation, it emerged that one of the contributing factors to the incident was the maps used by the air crew, which didn't show the location of the ski resort or the cables. In the chain of responsibility, who was at fault here? The data collectors for failing to record the detail about the ski resort and infrastructure? The data providers for failing to adequately document how the data should be used? The end users for failing to ensure that the spatial data used showed the appropriate detail for their particular purpose?

Will it ever be possible to guard against this type of inappropriate use of data? It is one thing to make volumes of data available on a website, but quite another to control how those data should be used. As more and more data are available, examples of misuse like this will become more common unless steps are taken, and soon, to provide rigorous guidelines for end users.

Almost everyone uses spatial data to a certain extent—travelers working out the best route from one location to another, visitors locating nearby attractions and services, planners providing details of proposed development plans, media professionals reporting on an incident in a city center and wishing to provide location information for context. These are just some examples of the everyday use of spatial data. However, most end users are not GIS professionals, and they often make use of spatial data without necessarily appreciating the limitations. This may result in anything from inadvertent copyright infringement, to the far more serious consequences of data misuse that result in severe economic or environmental impact or loss of life.

Microsoft and its partners have been supplying operating systems and software for many years now; they learned through experience, and fear of prosecution, to be very explicit about how and where the software should be used. Here is an excerpt from one of Microsoft's End User's Licensing Agreements:

> "Note on Java Support. The OS Components may contain support for programs written in Java. Java technology is not fault tolerant and is not designed, manufactured, or intended for use or resale as on-line control equipment in hazardous environments requiring fail-safe performance, such as in the operation of nuclear facilities, aircraft navigation or communication systems, air traffic control, direct life support machines, or weapons systems, in which the failure of Java technology could lead directly to death, personal injury, or severe physical or environmental damage. Sun Microsystems, Inc. has contractually obligated Microsoft to make this disclaimer."

Although capturing and providing metadata are now enshrined in many organizations as part of the data collection and maintenance process, very few provide such prescriptive documentation which clearly identifies for which applications the data are suitable or, perhaps more importantly, unsuitable. In many ways, this is a much more difficult task than recording how and by whom a dataset was captured (the scale, format, and update lineage and so on). Those who collect and provide access to spatial data are generally clear when describing their own data. However, what is also required are explicit statements as to where those spatial data should and should not be used.

Some say that such statements would impair the creativity that comes with using GIS. Others say that such statements give too much power to data providers, as they could use them to prosecute those who use the data inappropriately. To compound the problem, the use of spatial data has changed almost beyond recognition from the early city planning and cadastral applications to location-based applications in just about every conceivable industry and aspect of the economy. It would now be almost impossible to state what problems the data can and cannot be applied to. Still, many maintain that data providers should follow the lead of companies like Microsoft and provide statements highlighting inappropriate uses for the data.

Unfortunately, the current data access policies of many spatial data providers tend to avoid applying any such specific restrictions on the use of the spatial data and absolve the provider from any responsibility. Many do not provide explicit statements on inappropriate use. For example, the following is from the data policy for the CIESIN World Data Center we discussed earlier:

> "CIESIN offers unrestricted access and use of data without charge, unless specified in the documentation for particular data. Data are freely distributable and redistributable from CIESIN unless otherwise specified in the documentation accompanying particular data."

READ AND RESPOND

Examine the Esri Map Books website (http://www.esri.com/mapmuseum/index.html), an annual collection of some of the best GIS map output from users in government, industry, academia, and nonprofit organizations all around the world. In light of our current discussion, ask yourself the following questions about each:

What restrictions do you think operate on each of the datasets that are being used to create these maps?

How much of the content do you believe are public domain data?

Of the data used in these maps, how much do you estimate are free of charge, and how much had to be purchased?

Some data on these portals are available under specific licensing agreements. For example, the Global Administrative Unit Layers (GAUL) provides spatial information on administrative units for all countries in the world. The primary purpose for the data is "for the benefit of the United Nations and other authorized international and national institutions/agencies." However, for noncommercial purposes, the data may be used by other organizations and individuals. The license agreement states,

"The Food and Agriculture Organization of the United Nations (FAO) hereby grants you a license to use, download, and print the materials contained in the GAUL dataset solely for your own non-commercial purposes and in accordance with the conditions listed below. Copyrighted data made available to you through the GAUL dataset is subject to the conditions specified in the metadata. The licensee will at all times use due diligence to safeguard and protect all such confidential and proprietary information pertaining to the Licensed Data. Except as may otherwise be expressly permitted by applicable law, all redistribution to non-authorized users or commercial exploitation of any material contained in the GAUL dataset is strictly prohibited without the prior written consent of FAO. Any request for permission to reproduce or otherwise use any material contained in or otherwise made available to you through the GAUL dataset other than as provided for herein, should be directed to FAO."

Therefore, as the data user, not only do you need to have a clear purpose as to how you wish to use spatial data, but you need to understand, by reading all documentation and use statements, whether you are permitted to use the data you are accessing.

Data ethics and data politics

Sharing data is not new. People and societies have been sharing data for thousands of years and, over the generations, protocols and accepted practice have emerged governing the exchange of those data. Sharing digital data is, however, relatively new and we are still at a fairly naïve stage where we mostly believe all the advantages of the digital age—speed and ease of access, rate, and volume of data transfer—can only be for the good. The ability to share digital spatial data so easily and quickly represents a double-edged sword. On one hand, sharing digital data offers great benefit and potential to society; on the other hand, as with nuclear power, the Human Genome Project, advances in genetic medicine, and the web itself, this technology is open to abuse.

The advent of the web has revolutionized access to data and information. For many people, the comparative ease with which spatial data may now be accessed provides a previously unimaginable information resource that is put to very positive use. For others, that same resource provides another opportunity to oppress or harm people. Events in Mumbai in 2008 highlighted the potential for harm when it emerged that the terrorist attack on that city had been planned and coordinated using high-resolution spatial imagery and infrastructure layers provided by Google Earth. Despite subsequent attempts to have that particular supply of spatial data banned in India, the Indian authorities appreciated the benefit that such information could provide and are currently developing their own geoportal, Bhuvan, to "explore and discover virtual earth in 3D space with specific emphasis on Indian region."

International spatial data on land use, borders, and the location of and access to mineral rights, for example, may also be used for economic advantage, with some countries using it to pursue opportunities that are in their national interest. Such unilateral behavior may be to the disadvantage of neighboring countries who may view such activity as a threat. A recent, highly publicized case concerned the energy reserves in the Arctic Ocean. As the ice caps melt, countries are mapping their claims to the energy resources revealed and the stakes are high; it is estimated that nearly a quarter of the world's undiscovered oil and gas reserves are to be found there (Funk 2009).

For the data producers, some issues that are trivial or easily resolved within the administrative confines of a country can become much larger issues when they transcend national boundaries. Even the depiction of those national boundaries themselves can result in conflict where the boundaries are in dispute. For example, during 2010, the Indian government objected to a Google map showing land that India considers Pakistan-occupied Kashmir as Pakistani territory. According to Sachin Pilot, the Indian minister for telecom and information technology, "Any wrongful depiction of Indian map and its boundaries is liable for action under the India Information Technology Act" (*United Press International* 2010). In the end, Google agreed to adjust the map of the Kashmir region, which is claimed by both India and Pakistan and has been partitioned for decades.

The proliferation of web mapping services illustrates another international issue. In China, for example, foreign firms are not allowed to provide surveying and mapping services. Any activities on their behalf must be as joint ventures, or in partnership, with domestic Chinese firms (*China International Business* 2010). This is to allow authorities to check, among other things, that maps are labeled in accordance with Chinese rules and that sensitive information is removed. How can the international community work together when not only the workflows vary widely, but business practices do as well?

Summary and looking ahead

In this chapter, we have examined some of the initiatives behind the international collaboration for spatial data infrastructures. We have explored the benefits that such collaboration brings with regard to spatial data's production and dissemination and also some of the challenges that exist. We have also studied some of the data portals that have resulted from international collaboration and have contrasted these portals with the NSDIs. We have seen the different capabilities, accessibility, and licensing arrangements among different data portals.

In the next chapter, we will look at some of the other issues you should take into account when planning your GIS project. Although the data you use to address spatial problems is one of the most important considerations for your project, it is not the only one. Other factors include an organized plan, software, personnel, and other practical considerations. We will provide case studies as examples of projects whose organizers took these considerations into account.

References

Barry, Tom. 2008. "The Arctic Council—Approach to Spatial Data?" *Conservation of Arctic Flora and Fauna* (November). http://arcticportal.org/uploads/6Z/fY/6ZfY9ttS-bgOiDGw4X3kOw/The-Arctic-Council---Approach-to-Spatial-Data.pdf.

Bregt, Arnold, Joep Crompvoets, Erik De Man, and Lukasz Grus. 2009. "Challenges in Spatial Data Infrastructure Research: A Role for Transdisciplinarity?" Proceedings, 2009 GSDI 11 World Conference. http://www.gsdi.org/gsdiconf/gsdi11/papers/pdf/134.pdf.

Cohen, Sebastion. 2010. "Digital Treasure Maps." *China International Business.* (August 17). http://www.cibmagazine.com.cn/Features/Infotech.asp?id=1353&digital_treasure_maps.html.

Craglia, Max, Michael F. Goodchild, Alessandro Annoni, G. Gilberto Camara, Michael Gould, Werner Kuhn, David Mark, Ian Masser, David Maguire, Steve Liang, and Ed Parsons. 2008. "Next—Generation Digital Earth. A Position Paper from the Vespucci Initiative for the Advancement of Geographic Information Science." *International Journal of Spatial Data Infrastructures Research* 3: 146–167.

Crompvoets, Joep., Arnold Bregt, Abbas Rajabifard, and Ian Williamson. 2004. "Assessing the Worldwide Developments of National Spatial Data Clearinghouses." *International Journal of Geographic Information Science* 18 (October–November) (7): 665–689.

Funk, McKenzie. 2009. "Healy Mapping Mission." *National Geographic* 215 (May): (5). http://ngm.nationalgeographic.com/2009/05/healy/funk-text.

"Google Map of Kashmir Annoys India." 2010. *United Press International.* (August 21). http://www.upi.com/Top_News/International/2010/08/21/Google-map-of-Kashmir-annoys-India/UPI-46191282441812/.

Indigenous Mapping Network. 2009. "Conservation Geoportal Decommissioned." http://indigenousmapping.net/index.html.

INSPIRE. 2003. "Report on the Feedback of the Internet Consultation on a Forthcoming EU Initiative Establishing a Framework for the Creation of an Infrastructure for Spatial Information in Europe." (August 28). http://inspire.jrc.ec.europa.eu/reports/analysis_consultation_01092003.pdf.

Koerten, Henk. 2008. "Assessing Organizational Aspects of SDIs." In Crompvoets, J. A. Rajabifard, B. Van Loenen, and T. D. Fernández (Eds.) *A Multi-view Framework to Assess Spatial Data Infrastructures.* Melbourne: The Melbourne University Press, pp. 235–254.

Kuhn, Werner. 2003. "Semantic Reference Systems." *International Journal of Geographic Information Science* 17 (July–August) (5): 405–409.

Masser, Ian. 2005. *GIS Worlds: Creating Spatial Data Infrastructures.* Esri Press, Redlands CA.

Stevens, Alan R., Harlan J. Onsrud, and Mukund Rao. 2005. "Global Spatial Data Infrastructure (GSDI): Encouraging SDI Development Internationally." Conference proceedings, ISPRS Workshop on Service and Application of Spatial Data Infrastructure XXXVI (4/W6), Hangzhou, China, October 14–16.

Verduyn, Monique. 2009. "Spatial Data for Africa." *Geomatics International Magazine* 23(9) (September). http://www.gim-international.com/issues/articles/id1419-Spatial_Data_for_Africa.html.

7

PUTTING PUBLIC DOMAIN SPATIAL DATA TO WORK

Introduction

In the first six chapters of this book, we discussed the types of spatial data that are available and where to access them. We examined some of the issues surrounding spatial data in the United States and internationally, including free versus fee-based data, copyright, assessing data requirements, and recent developments surrounding national and international spatial data infrastructures and portals. Through a series of practical examples, you were able to experience firsthand what can be achieved using public domain spatial data and GIS tools.

In this chapter, we will take a look at some of the other components of GIS-based projects that you must consider when addressing spatial problems. GIS-based projects is a book topic in its own right and one that has been covered by other authors in much greater detail than we will be able to cover here. However, with this overview of project planning—how to choose the right data, the software and toolsets that are available, ethical concerns, and how to measure success—we want to encourage you to pay close attention to each of these considerations when working with spatial data. Knowing how to use data and understanding data is not enough; you need to consider how data can be used effectively to make decisions. To illustrate the potential benefits of using public domain data, we will review some case studies from organizations that have successfully incorporated public domain spatial data into a specific project or task.

Planning your project

Like any IT-driven project, a successful GIS-based project requires careful planning from start to finish. With the current abundance of spatial data, and the accessibility, ease of use, and wide variety of GIS tools available, the temptation has never been greater to start with the data and quickly generate some output. However, leaping straight into data acquisition and then retrofitting the problem to suit the data can lead to erroneous conclusions. This may result in the loss of support from those who had previously backed and promoted the project and, ultimately, the withdrawal of funding.

All too many GIS-based projects fail to meet their objectives because of insufficient planning and a failure to understand the nature of the problem to be solved. To plan a successful project, start by asking questions to establish exactly what the objectives are. Regardless of whether your project involves health, city planning, or business concerns, as you adopt a geographic perspective and engage in spatial analysis, the geographic inquiry process illustrated on the next page should help structure your project.

The geographic inquiry process begins by asking a geographic question. For example, a truck overturns on a bridge and dumps hazardous chemicals into a river. The first question to ask would be

Which community water supplies downstream would be affected?

A well-framed geographic question typically spawns related questions. For example, in the chemical spill example above, these may include the following:

How long would it take for the chemicals to reach these communities?
Which municipal water treatment facilities could be affected, and where do people live who use water from those facilities?
Which water wells could be affected, and where do people live who use water from those wells?

The Geographic Inquiry Process

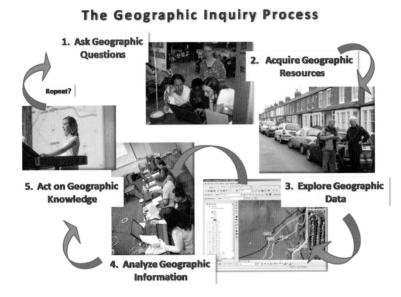

The geographic inquiry process is fundamental to solving problems with public domain data and GIS.

Which fields are irrigated with water from the affected streams?

Where should mitigation efforts be concentrated?

How could mitigation efforts most quickly and efficiently reach the affected rivers?

The geographic question should direct the types of data that must be gathered, explored, and analyzed to most effectively lead to sound decision making. The decision making is the action taken as a result of the inquiry. Most geographic inquiry is iterative; in other words, it leads to additional questions which, in turn, leads to additional data requirements and analysis. In the river chemical spill example, the project leaders may discover that they need data on small forest service roads, water wells, and municipal water facilities. This may result in the original question being modified or qualified in terms of scale, theme, or geographic location studied. For example, initial mitigation of the spill occurs at an extremely local level—extracting the truck from the water, sealing the truck, and removing it to a facility where the materials can be properly handled. Ongoing mitigation may cover an area dozens or even hundreds of kilometers along the river and in adjacent fields and towns. Organizations mobilized for the chemical spill may want to apply lessons learned to other types of chemicals or other types of terrain. Because spatial phenomena occur not only within space but over time, *when* the event occurs may affect the data, analysis, and decisions required.

For example, how would a significant rain event at the time of the chemical spill affect the situation?

It is important to note that the inquiry process reaffirms a point made earlier in this chapter: that the geographic question should influence the choice of data gathered and analyzed, and not the other way around. Too often, a dataset is selected and then a geographic question is posed simply because the data are ready and available. This can lead to ineffective decisions because the data are not appropriate to the question.

The best dataset for the project

Tomlinson (2003), in defining system scope, discusses the importance of understanding the nature of the information products that will be delivered at the end of a project. What input data will result in the best information product you can produce?

A key element in the planning stage for any GIS project is identifying the most appropriate data sources for your project. The increasingly diverse range of publicly available spatial datasets provides project managers, analysts, and end-users with an unparalleled repository of data resources to choose from. However, as we have seen in previous chapters, you need to choose carefully; you need to be certain about the source, update lineage, and any restrictions on use before considering using those data in your project.

We have already covered a number of these issues, but to recap, some of the important questions to ask of the data are the following:

- Who compiled the data? Would you consider the originator of the data an authoritative source? If not, is the source, such as volunteered geographic information (VGI), appropriate for your project?
- How were the data compiled? Were the methods used acceptable? If they were converted from another format, were any data lost or redefined during that process? For example, every time data are reprojected, they may not be as spatially accurate as in the previous iteration. Does this pose a problem for your project?
- Why were the data compiled? Were they assembled with one specific purpose in mind or will they support a number of applications? Is the reason, if any, is provided, sufficient for you to decide whether or not to use the data?
- In what medium were the data originally produced? Did the data come from analog sources, such as paper or film, and were these original sources cartographic or geographic in nature?
- Are the data available in a format which is easy for you to use in your project?

- If you are hoping to use the resources of a data portal, is the portal available in your native language or a language that you understand? If not, can you reasonably work with it anyway or will you need to have some, if not all, of the text of the portal translated? Are the data, including the attributes, in a language that you understand?
- Were the data captured at an appropriate scale? What was the density of observations used to compile them?
- What projection, coordinate system, and datum were used? Will these parameters be something that you can work with or reproject into something different?
- Are the data current? Do they need to be current?
- How accurate are the positional accuracy and the attributes?
- How were the data checked? Who checked the data?
- What are the restrictions on the data use? Do you have permission to use the data within your organization? How are you permitted to use the data?
- Are the data complete? Are there any features or layers missing?
- What is the areal extent of the data?
- What are the attributes of the data, and are they understandable and complete?
- Do the data include metadata? If not, be very wary of using the data. If metadata exist, have they been updated to reflect the current condition of the data?
- Finally, and perhaps most appropriate, are the data relevant to your project?

It's not enough to simply check whether metadata are present or not. Much more important is who compiled the metadata and how well the metadata truly represent the data quality. With the ease of filling out metadata forms nowadays, the temptation for data providers is to copy metadata content from one dataset to another, without customizing the metadata to the individual dataset being described.

One thing should be clear from this book, and is worth repeating, that despite the plethora of data repositories now available—and we have discussed some very successful portals in earlier chapters—not all are created equal. Some portals are easier to use and better stocked than others, providing an invaluable resource for GIS practitioners. Some promise much but deliver little, providing descriptions of data but no link to access the data. Other portals are more like community network sites or simply exist to disseminate news about a data initiative. One such example is Data Observation Network for Earth (DataONE), which organizes data for environmental observation but, as of yet, offers little in the way of spatial data for download and further analysis. On the other hand, some online resources that sound similar to others or appear unremarkable, such as the North American Environmental Atlas discussed earlier, are a gold mine of data and offer an abundance of data formats and themes and a variety of means to access the data. Some long-established data repositories such as the US Geological Survey (USGS) National Map offer a wealth of data but the reliability and downloading mechanisms remain a source of frustration for most users. Fortunately, the proliferation of online data resources usually means the required dataset may sometimes be obtained from a more user-friendly service provided by another agency, such as the US Department of Agriculture (USDA) Geospatial Data Gateway.

Some sites, such as GapMinder, provide useful resources (for instance, data visualization tools) but do not allow the user to access spatial data directly. Still, they are worth investigating, as some of the data may be downloaded. For example, Excel spreadsheets on the WorldMapper cartogram visualization website contain country codes that are easily joined to a country map in a GIS package.

Another important consideration in gathering data, which seems almost too obvious to state but is important nonetheless, is knowing when to quit. At what moment do you, as the analyst, decide you have enough data to meet your requirements? That will partly depend on the geographic inquiry process discussed earlier and how many other questions arose as you established the parameters of your project. In the past, for many GIS-based projects the data-gathering stage tended to expand way beyond the original estimates.

Tomlinson (2003) noted within a "typical municipal GIS, data entry can consume 80 percent of the time it takes to implement a GIS project . . .." This left comparatively little time for analysis and assessment—time and budgetary constraints often meant projects were concluded before all options had been investigated.

Today, it is easier to locate and get access to data, but because of this, you need to be disciplined and use only those datasets that are relevant to your project. If you don't, you will be no better off than GIS practitioners in the early days, as you too will find that only 20 percent, or less, of your project time is left to analyze the data and make decisions based on your analysis. Say you are planning to use a digital elevation model (DEM) in your project and you find an online resource that provides both 10-meter and 30-meter DEMs. If the 30-meter resolution DEM is more appropriate given the requirements of your project and the extent of your study area, don't be tempted into getting the higher resolution DEM just because it is available. Similarly, if real time data are available but your project doesn't specifically require them, use last year's or last month's data if that's more appropriate.

For each dataset you download, you need to store it, process it, understand it, manipulate it, and develop it into a useful information product. Although there is a strong temptation to hoard more data that you really need "just in case," learn to say "no." Your project will benefit in the long term as you achieve a better balance between data acquisition and data analysis.

Choosing the right software package

Along with choosing the most appropriate dataset(s) for your project, your choice of software to process those data can have a significant impact on the cost of the project, the analytical capabilities you will have access to, and any support or custom development you may require to complete the project.

For many, the first thing to decide is whether to pay up-front for proprietary software packages and services, such as the ArcGIS product suite, or use the increasingly available, free Open Source Software (OSS) GIS packages, such as GRASS, Quantum GIS, OpenMap, and MapWindow.

OSS LICENSING

The license agreement for an OSS GIS package allows end users to operate the software for any purpose, to modify it as required, and to redistribute copies of either the original or modified program without having to pay royalties, comply with copyright restrictions, or pay an annual maintenance fee.

Free, OSS, and proprietary GIS software are very different in terms of up-front and ongoing costs, risk, and available support. It may seem that getting GIS software for free is the obvious choice, but free doesn't mean there are no costs associated with it. Paying for proprietary software could save you time, and ultimately money, in the long run. Some points to consider include the following:

- Is the software available for your organization's chosen operating system?
- Do you or your organization have the skills to fix bugs?
- Do you or your organization have the resources to support a number of users who may require access to the software?
- Does the OSS GIS package you are considering provide the necessary analytical tools to enable you to produce the output your project requires?

- Do you or your organization have the resources to customize the software to meet specific requirements?
- If your project demands a more tailored solution, will you be able to develop the application you require?
- Is your project mission-critical or are you aiming to provide essential services? If so, what guarantees and support does your choice of GIS software provide?

Whether you choose to go for proprietary or OSS GIS software, you should be aware of the advantages and disadvantages offered by both. An emerging trend in many organizations appears to be a mixed implementation; OSS desktop applications and web implementations in conjunction with proprietary enterprise spatial data management solutions.

Other important considerations influencing your choice of software are the dataset support, mapping tools, and analytical tools your project requires. Whether you decide to go the proprietary or OSS route, choose a software toolset that meets all or as many of your requirements as possible. The right toolset will

- allow you to work with the data sources that are relevant to your project, for example KML, OpenStreetMap, shapefiles, and CAD drawings;
- provide the appropriate level of display and analytical tools, whether such basic viewing and query tools as provided by ArcGIS Explorer or, the multi-user support, advanced analysis, and processing capabilities of an enterprise solution; and
- support the output formats necessary to generate the information products you are required to produce.

Off-the-shelf map templates and data

To help reduce the amount of time and effort required to set-up and configure their GIS application, many users are turning to map templates and pre-bundled basemaps to provide the basis for their projects. The appeal for many is the ease of use, the integrated reference mapping, and pre-defined display styles. Templates are a great way to apply a consistent approach to a production process, as they can be easily shared within an organization.

Although templates can be a time-saver due to their ease of use, that same advantage can occasionally result in inappropriate or misuse of the template. There is often a temptation to mold the project to the template and the supplied basemaps and accept the data in the format and style supplied, even though the specifications in the project plan in terms of scale, currency symbology, and so on are different. If the template

TEMPLATES SAVE TIME

If you're thinking of using a template in your project, have a look at the ArcGIS Resource Center and the Map Templates section. There, you'll probably find a template that could save you some time. If you have your own template, why not consider sharing it with other users and upload it to the Map Template Gallery?

doesn't quite meet the project specification, then any analysis performed on the data, or output derived from the data, may fail to meet requirements. The project could be considered a failure and the perceived benefit of GIS questioned.

Map templates and basemap bundles have a role to play within many successful GIS projects, but think carefully about exactly what is being provided. If the data supplied aren't quite right, find an alternate source. If the display and symbology aren't exactly as you require, spend the time to adapt the template to suit your needs.

We will return the topic of map templates in chapter 9 when we look at cloud computing.

Trained personnel

Arguably, one of the most important components of any GIS project is the people who are trained to use the chosen GIS software and undertake the spatial analysis required. Some of the staffing functions that may be required on many GIS-based projects include program managers, GIS analysts, system administrators, spatial database specialists, GIS technicians, and end users. In practice, individual staff members may perform one or more of these functions. As GIS increasingly moves onto mobile devices and the Internet, mobile and web experts are becoming essential.

During the planning stage, you should consider, given the requirements of the project, what skills exist in-house already and where they can be best applied. If any skill shortages are identified, can those skills be acquired via training or is hiring contractors or recruiting permanent staff an option? Many organizations successfully contract out some of the more specialist aspects of a project, for example data conversion and infrastructure maintenance.

You may also wish to consider using external GIS consultants to help with requirements gathering, project planning, or to provide a customized solution for your project. Using consultants can help establish GIS projects quickly and effectively, with best practices incorporated into the project from the outset. However, too much reliance on external help can be detrimental if there is little opportunity for knowledge and skills transfer to the permanent staff that must sustain the project for months or years, long after the consultants have gone. The key is finding the right balance for your organization and your project.

For companies and organizations that have adopted or are in the process of adopting GIS as a mainstream element of their IT infrastructure, training is vital. Training in the use of application and database software, programming skills, system administration and design, support, IT project management, and so on, should all be considered where appropriate. Many projects fail because of inadequate or inappropriate training. If staff is not able to use the technology to its best advantage, the organization will never realize the benefits of GIS.

Measuring success

GIS is a tried and tested technology. There are numerous examples throughout the world illustrating the benefits of using this technology to process and analyze spatial data to solve some of the world's most pressing environmental, economic, and social problems. However, although GIS projects are by nature technology based, simply having the latest version of software or the latest must-have piece of hardware, does not guarantee success. A project will only be considered successful if it benefits the individual or organization undertaking the work.

Measuring that success involves identifying and quantifying the perceived benefits of the stated approach. Key project indicators, such as cost- or time-savings, will confirm if project goals have been attained.

In the context of solving spatial problems with public domain data, a look at some of the indicators of success that specifically relate to the data should help illustrate the importance of considering the use of these types of data.

FOR FURTHER READING

Establishing the business case and measuring the success of a project is an extensive topic in its own right and beyond the scope of this publication. However, if you are interested in reading more about the business case for GIS and return on investment, have a look at the *The Business Benefits of GIS: An ROI Approach* (Esri Press 2008) by David Maguire, Victoria Kouyoumjian, and Ross Smith.

Will public domain spatial data save you money?

For many organizations, legitimate concerns about copyright, legal responsibility, currency, and unrestricted access to the data, compounded by a lack of communication within and between organizations, have resulted in costly data capture exercises to recreate datasets that may already exist elsewhere. As we saw in chapter 5, the problem of dataset duplication is not uncommon. Although capturing your own data does mean each dataset is highly customized to the requirements of the project, with no royalty fees and no (or fewer) restrictions on use, data capture is expensive. For many projects, data capture is probably the single largest expenditure. Given the considerable costs involved, reusing existing data should be one of the first considerations during the planning stage of any project. Questions to consider:

- Is commercial sensitivity an issue for your organization?
- How much data do you need? If you need access to large volumes of data, can you afford to create it all?
- Can you afford the ongoing costs of data storage, update, and maintenance if you choose to collect the data yourself or could you take advantage of online Data-as-a-Service and data streaming technologies to reuse existing data?
- Are other groups or organizations trying to achieve the same goals you are? Could you share data resources?

Using spatial data that already exist in the public domain can significantly reduce project costs. Even if you must reformat the data and perform some data manipulation before it will meet your particular requirements, reusing existing public domain data may still save you money in the long term. Don't embark on a costly data capture exercise unless you are convinced it is the only option for your project.

Will public domain spatial data save you time?

As data collection and data preparation can consume as much as 80 percent of the time and resources allocated to a project, any opportunities for reducing the amount of time spent on these activities should be considered. Some questions to pose when assessing the data requirements of your project:

- Will using public domain spatial data allow you to meet deadlines and result in the information products being delivered on time?
- Will using public domain spatial data reduce the amount of time you or your staff will have to spend simply getting the data into a suitable format for the project?
- Will you be able to use project resources more effectively as a result?

Using public domain spatial data can result in considerable time-savings, but they have to be the right data for the project and they have to be available when needed. Otherwise, the potential benefits are lost.

An increasing number of data portals are now available and provide access to public domain spatial data, and the technology behind data access and delivery has improved significantly in recent years. If you are considering using public domain spatial data, you should factor in some time spent trying to locate the right dataset. As Tomlinson (2003) also noted, sometimes knowing where to look and how to search is a challenge in itself. You may, of course, be lucky; your first inquiry could turn up a dataset that meets your requirements. However, more often than not, you will spend time visiting various portals and trying to find the dataset(s) that you can use in your project.

Some practical considerations in obtaining spatial data

There are many issues to consider when you opt to use public domain spatial data:

- Will you need an account to access the data?
- Will you need to do something to your web browser to access the data, such as add the site to your trusted list, turn off pop-up blockers, or add a Silverlight or other plug-in?

- Does the site hosting the data contain annoying or malicious pop-ups, advertisements, or something else that could make the process of accessing the data inconvenient, or worse, harm your computer?
- Is the site available during the time and time zone when you need it, or is it only available periodically?
- Does the site provide a rapid pipeline to download data at a reasonable speed, or does it require too much of your project time to access the data? Does the site (such as the GIS Data Depot from GeoCommunity) provide two different means of obtaining the data—a free but slow method and a fee-based faster method? If you need to save time and money, consider paying the fee.
- Is there a maximum number of files that you are permitted to download? Is there a maximum amount of data in terms of megabytes or gigabytes that you are permitted to download? If so, will these limits cause a problem?
- Do you need to actually download the spatial data for your project to your own computer, or can you stream the data from a mapping service?
- If you are streaming data for your project, will your project suffer if there will be times when you will not be able to access the Internet?
- If you download data that have been compressed to reduce transfer times, do you have the necessary software to uncompress it?
- If you download the data, does the system allow you to download multiple tiles or cells at once, or must you access the site multiple times for multiple datasets?
- If you download the data, do you know where the data will be saved? Can you override the default data folder and specify your own folder?
- Do you download the data by outlining a box on a map, by clicking map cells, or by visiting an FTP site? Does the site contain an alternative way to download the data in case the primary method is slow or not operational?
- Are the data on the site vertically integrated with one another? For example, have the contours been smoothed out over water bodies and do the census blocks nest inside the block groups?
- Will the file names indicate what you are receiving, or will you need a look up table?
- Where will you store the data, and do you have enough room on your computer for the data?
- How will you back up your data? Will the backup method be a problem?
- What format are the data in, and will this format work in the type of GIS software that you have?

- Will you be able to edit the data once you download them, or are the data read-only or otherwise locked?
- How do you cite the source you are obtaining the data from?
- Do you know enough about the projection, datum, resolution, and scale of the data to judge whether they are fit for your purpose? If you do have to project the data, will the projection cause errors that are unacceptable for your accuracy requirements?
- Do the data have the types and numbers of data fields in the attribute tables that you need? Will you know what each of the field names signify?

There are no definitive answers to all these questions; the response to each is unique to the application, organization, and analyst seeking to use the data. However, these are the types of questions that you need to ask again and again.

Setting up your project environment

Numerous best practices have been written for setting up your GIS project, but it still merits attention here, even before you begin looking for data. Here are a few points to consider:

- Who else will be working on the project? Do you have access to staff with the skill sets you will require?
- Does everyone on the project have access both to the software tools and the data?
- Do you need to grant different levels of access to different users (readers, editors, or data administrators)?
- Which datasets do you need to download, and which could you stream from the cloud?
- Do you have enough hard disk space to store the data if you download them?
- Is your naming convention consistent and easily understood for all files and folders in your project workspace?
- Does your software package provide configuration options? (If so, setting up your project preferences at the outset will save you time later on.)
- If you are basing your project on a template, does everyone have access to it?
- Will you need access to any peripheral devices, such as printers or scanners?

Time spent preparing for the project is time saved.

SETTING UP PREFERENCES

For ArcMap users, setting up project preferences includes setting up the Map Document Properties, environment variables, and the ArcMap Options. In Map Document Properties, fill in the title of your project, create an empty geodatabase into which you will load all of the data you will download and/or create. Typically, you will use a variety of file types and maintaining them in a single geodatabase will reduce your file management overhead and provide you with additional structure for the data. To prevent having to search your entire hard drive or networked drive, specify your default geodatabase so any data you may add or any results you may generate will always be saved here.

Next, check the box indicating that you will store relative path names to data sources; this avoids being tied to hard coded path names. In the Geoprocessing Environment, set your current and scratch workspace to your working folder and database. Under Geoprocessing Options, ensure that you "Overwrite the output of geoprocessing operations" and check "Add results of geoprocessing options to the display" so your results will be added to the table of contents.

Under the Customize menu, under ArcMap Options, visit the Raster tab and Raster Dataset. Be sure that the option to Use the world file to define coordinates of rasters is checked; if you have raster data, it will be registered to real-world coordinates. Under the Metadata tab, select the type of Metadata format you prefer. Under Customize, be sure that any extensions you want to use are enabled.

Lastly, save your map document—and while you are working, save your project often. Create a backup of your map document and your data layers in a different location or storage device. Consider backing up your data in the cloud as well as on physical media.

Finding public domain spatial data

Before the advent of robust search technologies, web pages containing links provided the early one-stops for spatial data by listing sources for a variety of themes. Many of these were assembled by universities and occasionally by nonprofit, government, or commercial organizations. Examples of these include Columbia University Libraries' Digital Social Science Center, Ohio State's Numerical Cartography Lab, and the University of

North Carolina's Spatial Data Online resource. Two of the first were the Alexandria Digital Library at the University of California, Santa Barbara, and the resource established in 2001 by the University of Arkansas, originally named Starting the Hunt. As the technology has developed, a number of purpose-built spatial data portals, such as the Geospatial One-Stop Geodata.gov and the European Commission's INSPIRE portal, were established and financed by government agencies and consortia.

Nowadays, most data users search for public domain spatial data in the same way they search for anything else on the web—using a search engine such as Google and Yahoo!. Although this method can provide quick and relevant results, there are some practical things you can do to maximize the efficiency and effectiveness of Internet searches. First, become familiar with the advanced settings on the search engine you are using; this can save you valuable time. Most have the capability of searching for all words, exact words, one or more specific words and searching within a specific site or domain. Search terms such as "GIS" and "spatial data" can be more effective if used in conjunction with the theme of data you are seeking and the scale or region, such as "global biomes GIS." Including the format of the data, such as the terms raster, vector, shapefile, geodatabase, KML, SHP, MXD, XLS, map package, layer package, SID, and GeoTIFF, in the search string can also help. Knowing how spatial data are stored and made available on the Internet provides more options for searching. For example, including the "allinurl" string will search for a specific path; if the string "allinurl: arcgis/rest/services" is searched, public Esri ArcGIS REST (representational state transfer) services can easily be discovered. These sites are running REST APIs, which provide open web interfaces to GIS services hosted by ArcGIS for Server. As all resources and operations supported via the REST API are accessible through a URL for each published service published, and they use a common path in the URL, they will be located with the "allinurl" search string.

FINDING HIDDEN DATA

Sometimes, data on the web exists but is hidden in historical pages. The Wayback Machine is a digital time capsule created by the nonprofit organization Internet Archive, which archives web pages from 1996 onwards. You may find the data you are seeking there.

With the plethora of web browsers and different versions—some running animations and other functions—if you experience trouble accessing any dataset, your difficulty might be due to the browser, rather than the site. Try another browser, or try an updated version of the same browser, and make sure that whatever plug-ins you need (such as Flash or Silverlight) are installed. There are also tools that allow you to download large numbers of files at one time, such as *wget* in the Ubuntu operating system or Firefox's *download all* add-on (with this tool you can download all the map layers from the national atlas of the United States at once). Utilities like these may be useful for the spatial data libraries you wish to access.

One of the problems plaguing most spatial data collections and portals is that they become difficult to maintain and keep up-to-date. A further complication is the continually changing nature of the web itself. For example, one of oldest and largest collections, Oddens, is no longer maintained; visitors to the site are advised that "This site is no longer maintained. Use 'as is' is still possible." Still, such collections should be considered when you are searching for data, because some of the individual entries in the collections are often missed by search engines. Collections like these are also particularly useful if they are maintained by GIS professionals rather than a general web developer. Data collections developed and supported by the GIS industry—such as the GIS Data Depot maintained by the GeoCommunity, Dundas Data Visualization Inc., and Esri's ArcGIS Online—generally stand a better chance of enduring as a long-term spatial data resource than grant-funded or ad-hoc academic projects, although some exceptions to the rule do exist.

GIS is an analytical tool used in many research studies throughout the world. It may well be worth the effort to investigate relevant research literature in your search for data, including journals, edited volumes, books, and even newsletters from major research organizations and professional societies. Your search in professional society literature should include the geospatial field, of course, such as the Association for Geographic Information, the Geospatial Information & Technology Association, the Urban and Regional Information Systems Association, the American Congress on Surveying and Mapping, and the American Society for Photogrammetry and Remote Sensing, to name a few. You should also examine those related to geography, such as the International Geographical Union, the Royal Geographical Society, and the Association of American Geographers. Many organizations and GIS software companies host annual conferences, including the Esri International User Conference, GeoInt (GeoIntelligence), and GeoWeb. Many of the proceedings from those conferences are available online.

Since GIS is also applied in many disciplines, consider the professional societies and literature tied to these disciplines, such as human health, business, engineering, transportation, energy, public safety, law enforcement, environmental studies, hydrology, climate, and biology. Your search should include major publishers of GIS-related textbooks and journals, such as Taylor & Francis, Wiley, and Esri Press, and also online indexing services, such as university-hosted bibliographies and Google Scholar. These resources may highlight a study that is investigating the same issues that you are addressing and you may find that the author(s) of the study is willing to share data with you or point you to a relevant data source.

Remember, even in the twenty-first century, despite all the online portals, clearinghouses, and server farms, not all geospatial data are online. A vast amount still resides on individual workstations, local intranets, and even CD-ROMs and other physical media. In many cases, the organizations or individuals have neither the staff nor the time to place the data online or cannot upload it because the data are restricted in some way. An often overlooked but effective way of obtaining data is the more traditional direct approach of contacting the data provider directly via e-mail or telephone. It may save you precious project hours or even days.

Social media resources are often more up-to-date than web directories. For example, blogs provide an excellent way of keeping in touch with new datasets posted online. Some bloggers, such as the OpenData Free GIS Data Tips, Publications and Open Resources blog and the Free GIS, Remote Sensing, Spatial, and Hydrology Data blog, focus on free spatial data. It may also be worthwhile following blog authors on Twitter, Facebook, Google +, or LinkedIn to stay up-to-date with the very latest information and technology updates.

Some organizations have Facebook or LinkedIn pages that provide regular updates on new data resources. In contrast to the often inadequately monitored contact facilities offered by many websites, social media tools have the added benefit of allowing you to send a message to the page administrators, or even the authors directly, to inquire about a particular posting or spatial data resource. As these pages are updated frequently, you have a much greater chance of getting a response to your inquiry.

Other considerations

Like many other digital technologies, GIS developed rapidly and became established before concerns were raised about the ethical implications of its use. As a user of GIS, you are responsible for the ethical use of the information and the conclusions you derive:

Are you propagating any errors in the data?

Are you protecting people's privacy in the data you are using?

Are you creating maps that are clear and unambiguous and will not mislead your readers?

Are you disclosing your data sources?

Are you being fair in your assessment of the problem?

Are you respecting people's lives, rights, and property?

In chapter 4, we noted that spatial data are a means to an end, not an end in themselves, in that data provide the basis for analysis and generate the information for decision-making. If the data you choose are inappropriate or inaccurate, the value of anything you derive from those data will always be tarnished.

FOR FURTHER READING

To learn more about planning GIS projects, an excellent place to start is the book: *Thinking about GIS: Geographic Information System Planning for Managers* (Esri Press 2011) by Roger Tomlinson, a key founder of GIS. A companion book, *Building a GIS: System Architecture Design Strategies for Managers* (Esri Press 2012) includes a Capacity Planning Tool that guides the process of deploying and management of a GIS. The book *Mapping Global Cities: GIS Methods in Urban Analysis* (Esri Press 2006) illustrates how GIS can be used to plan and sustain urban areas. *The GIS Management Handbook* (Kessey Dewitt Publications in association with URISA 2009) from the Urban and Regional Information Systems Association provides information on how to develop, implement, and manage GIS programs and projects, aimed at managers.

Case studies

The exercises that accompany this book allow you to see for yourself what can be achieved using public domain spatial data. They are not just academic exercises but, as the following case studies also illustrate, public domain spatial data are being used every day in many organizations throughout the world to help solve real problems.

Case study 1: SEPA

The Scottish Environment Protection Agency (SEPA) is Scotland's public sector environmental regulator, whose "main role is to protect and improve the environment" through the regulation of activities that cause pollution and by monitoring air, water, and land quality. Part of that charge includes the responsibility to produce River Basin Management Plans (RBMP) for Scotland.

River basins include all the rivers, *lochs* (lakes), wetlands, and groundwater that eventually drain into the sea, as well as estuaries and adjacent coastal waters. Taking a source-to-sea approach is considered essential for the effective protection and improvement of the water environment, as something which affects one part of a river basin will often have an impact elsewhere in the basin. RBMPs ensure that public sector bodies, businesses, and individuals work together to protect the water environment by coordinating all aspects of water management.

To help illustrate the RBMPs, SEPA has produced a RBMP interactive map that is available online. Many of the water body features, including coastline and loch boundaries, have been derived from the contour data in the Ordnance Survey Land-Form PANORAMA dataset, recently made available for free under the OS OpenData program. The catchment area boundaries have been derived from the gridded digital terrain model (DTM) version of the same dataset.

Some of the supporting river catchment geology characterization work was based on the small-scale bedrock and drift (superficial deposits) data from the British Geological Survey (BGS)—the 1:625,000-scale Digital Geological Map of Great Britain (DiGMapGB-625). Subject to acknowledgment of source, the BGS data are now available at no cost for commercial and public use.

The revised licensing restrictions of the OS OpenData program and the unrestricted access to the BGS small-scale data mean that SEPA is now able to make more of its own derived data available in the public domain and is now investigating options for visitors to the SEPA website to download the data.

Case study 2: OpenStreetMap

OpenStreetMap is a collaborative, or crowdsourced, project which aims to create a free, editable map of the world and provide free geographic data for anyone who wishes to use them (Haklay and Weber 2010). The project was a direct response to the prohibitively restrictive, legal, and technical barriers to working with spatial data in the United Kingdom.

Started in 2004, the initial map was compiled by volunteers completing ground surveys with handheld GPS units. The data were then stored in the OpenStreetMap database where they were edited to correct errors and augment with additional features. As the project has grown, it soon became apparent that additional data sources would be required to speed up the rate of data capture and continually improve data quality. As OpenStreetMap was founded on the principle of providing free spatial data, any potential data sources that could be incorporated into the OpenStreetMap database must be free data and available under licensing arrangements that are compatible with the OpenStreetMap Creative Commons Attribution-ShareAlike (CC BY-SA) license.

TERMS OF THE LICENSE

Remember from chapter 1 that the Creative Commons Attribution-ShareAlike license allows others to "distribute, remix, tweak, build on the work and use for commercial purposes as long as originator is given credit and all new creations are licensed under identical terms."

In practice, this means all contributing OpenStreetMap data sources must be out of copyright or in the public domain. Some of the government and commercial public domain data sources that have been incorporated into the OpenStreetMap database so far include the following:

- TIGER data from the United States
- Landsat 7 satellite imagery (and the derived Prototype Global Shoreline data)
- Aerial imagery from Yahoo!
- PhotoMaps, major Australian cities and some rural areas provided by NearMap Pty. Ltd.

OpenStreetMap offers free mapping services based on these public domain data sources that cover a variety of themes. Here are some of these services, and their coverage:

- OpenStreetMap—general map service, worldwide
- OpenCycleMap—cycling tours and routes, worldwide
- OSMTransport—public transport networks, worldwide
- OpenStreetBrowser—various categories available, including culture and religion, leisure and sport, Europe

We will return to OpenStreetMap in chapter 9 when we look at crowdsourced geographic data.

Case study 3: geoland2 environmental monitoring

In chapter 6, we mentioned Europe's Global Monitoring for Environment and Security (GMES), a program that collates data from a variety of earth observation sources (earth observation satellites, ground stations, airborne, and sea-borne sensors). As part of GMES, the geoland2 project takes the core and mapping data from the continental GMES core data services and develops derived mapping and information services and products (Meissl and Triebnig 2010).

One of those services, the Spatial Planning service, takes data from the GMES land monitoring service, which captures data on land use and land cover change, soil sealing, water quality and availability, spatial planning, forest monitoring, and global food security, and integrates those data with additional geospatial and statistical data into "geographical information procedures, toolsets, and models." This derived product supports the analysis of land use consumption and demographic changes across administrative boundaries, at regional, national, and European levels.

At the start of the book, we discussed the differences between data and information—data being the body of facts or figures, which have been gathered systematically for one or more specific purposes and information being data which have been processed into a form that is meaningful to a recipient and is of perceived value in current or prospective decision making. The geoland2 project provides a good illustration of the process of converting the raw public domain earth observation data into a spatial planning information product that

- provides spatial planning information with an improved spatial resolution and thematic scope;
- supports spatial planning authorities with consistent and comparable EO-based information products and tools to describe, explain, and forecast urban land use changes; and
- moves from observing and monitoring to the evaluation of policy options.

These spatial planning products and derived data products are also available in a variety of raster, vector, and alphanumeric formats, including Esri shapefiles, Esri rasters, and GeoTIFFs.

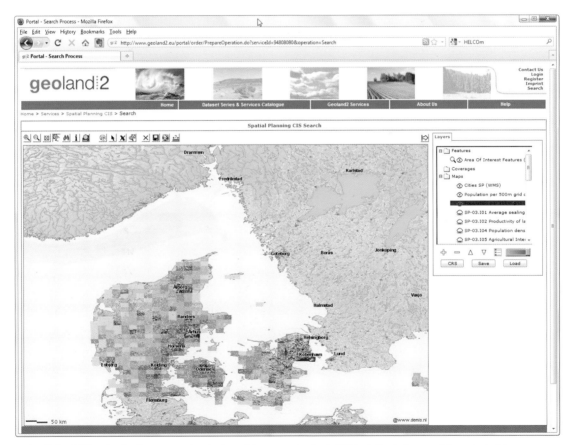

geoland2 Spatial Planning Service. Courtesy of geoland2, a collaborative project (2008-2012) funded by the European Union under the 7th Framework Programme (project number 218795).

Case study 4: WRI Forest Atlases

The World Resources Institute (WRI) is an independent, nonprofit organization established in 1982 to undertake research and provide solutions for some of the world's most serious environmental issues, such as deforestation, climate change, and natural resource management. As part of its focus on people and ecosystems, the WRI has developed a Forest Landscapes Initiative which aims to "increase the ability of governments, businesses, and civil society to *protect* intact forest landscapes, *manage* working forests more effectively, and *restore* deforested lands." To support this initiative, WRI has developed a series of Interactive Forest Atlases for countries in Central Africa, South America, Asia, and Europe (Archy 2011). These atlases provide a resource of publicly available forestry information to support national resource management and policy making in those countries.

The Central African Republic (CAR) is one such country which benefits from the WRI forestry initiative. Prior to 2007, insufficient forestry data were available for the government to adequately monitor and manage the densely forested southwestern corner of the country, an area on which both the local and the national economies were heavily reliant. Although, in theory, public forestry data were available in CAR prior to 2007, much of the data were virtually inaccessible, ensconced in both government and private sector data stores and not freely available. Any data that did exist were of marginal quality, available at different scales for different areas, coverage was often incomplete and with many overlapping and conflicting land use categorizations. Boundary definitions were also incorrectly interpreted. The slow rate of conversion from paper to digital data also hindered progress.

The CAR government recognized that to improve the management of this vital natural resource and ensure national and international compliance within their forestry industry, the government, private sector, and civil society needed access to current, accurate, and reliable spatial forestry data. A 2007 agreement between the CAR Ministry of Water, Forests, Hunting, and Fishing (MEFCP) and WRI launched a program of data acquisition and improvement. Working with a combination of the original legal definitions of the boundaries, existing spatial data, free public domain Landsat and ASTER imagery, and fee-based ALOS and DMCii imagery, one of the first tasks was to reinterpret national, regional, and individual land use parcel boundaries. Without legally binding and generally accepted boundary definitions, it would be impossible for the government to allocate and enforce logging permits and restrictions. WRI was also able to compile an integrated spatial dataset for the region which included the public road network, logging infrastructures, rivers, and protected areas.

The first release of the Interactive Forest Atlas for CAR was published in 2010 as a collection of interactive maps, datasets, and accompanying reports which were made available both as a CD-ROM and as downloadable data (in shapefile format) from the WRI website. The available themes include habitat, hydrography, administrative boundaries, and vegetation. The data are available under a Creative Commons Attribution Non-Commercial (BY-NC-SA) license, although the WRI acknowledges the difficulty of enforcing the noncommercial aspect of the licensing restrictions on subsequent reuse of the data.

Although the initial work focused on forests of the southwest corner of CAR, the government was quick to appreciate the benefits it could deliver through access to information on logging permits, accurate assessments of annual harvestable areas, and soon-to-be-available surveillance data to help identify any irregular activity. In neighboring Cameroon, a similar initiative to combine accurate forest use mapping with monitoring of logging activities via satellite imagery has played a significant role in helping to reduce levels of illegal logging by an estimated 50 percent over the past decade. Given the vital revenues lost through poorly monitored logging activities, the CAR government is understandably keen to see this initiative expanded to the rest of the country.

The work of the WRI and MEFCP in CAR—in this case, taking the original spatial data, augmenting and improving those data with other public domain spatial data sources, then providing a publicly available forest information resource that is available for both government and local communities—provides another excellent illustration of the many benefits that may be derived from working with, and providing access to, public domain data sources.

Case study 5: Acid mine drainage in the Animas River watershed, Colorado

Energy resources such as coal, oil, and natural gas have become the building blocks of modern economies. However, many economies are still grappling with the costs of exploiting these resources; they often contain harmful chemical substances that, when mobilized into air, water, or soil, can adversely impact human health and environmental quality. One of the side effects of mining activity has been the alteration of local water chemistry by releasing minerals into the water or as a result of the chemicals used and mining techniques adopted. A common problem is the decrease in the pH of water to the point of acidity. The USGS conducted scientific research to determine the chemical reactions that create acidity and which, in turn, led to the precipitation of dissolved metals. The acidity of coal-mine drainage is caused primarily by the oxidation of the mineral pyrite (FeS2), which is found in coal, coal overburden, and mine waste piles, also known as bings. The rate of pyrite oxidation depends on the reactive surface area of the pyrite, the oxygen concentration and pH of the water, the forms of pyrite, and the presence of iron-oxidizing bacteria. Despite improvements in methods of prediction and prevention, acid mine drainage problems persist in many areas of the world.

Passive and active systems have been developed to treat coal-mine drainage and raise the pH of the water to control the precipitation of dissolved metals. However, prevention is always better than cure; predicting and preventing acid mine drainage from occurring is preferable to having to perform remedial treatment after the problem has occurred. Static and kinetic chemical tests have been developed to aid in predicting potentially acidic drainages. A common technique is to determine the acid-base potential of the overburden, or rock strata, in the area. One of these chemical predictors is the neutralization potential (NP), which can be combined with other parameters such as maximum potential acidity, to provide a predictive guide as to whether acidic drainage is likely to occur.

To address the issue of human exposure to toxic substances derived from energy resources, the USGS Energy Resources Program developed a project titled "Impacts of Energy Resources on Human Health and Environmental Quality." With the aim of providing policymakers and the public with the scientific information needed to accurately measure the human health and environmental consequences of meeting energy needs, the project generated public domain spatial data for a series of GIS-based assessments. The project began during the late 1980s for a few watersheds, but with attention in 2010 on the spatial aspects of human health, the project is receiving renewed attention.

In the Animas River watershed in southwest Colorado, historical mining left a legacy of acid mine drainage and elevated concentrations of potentially toxic trace elements in surface streams. The spatial data generated for this project are in the area upstream from Silverton and include the Mineral Creek, Cement Creek, and upper Animas River basins. The study demonstrated how a watershed approach can be used to assess and rank mining-affected sites for possible cleanup (Buxton et al. 1997). The study was conducted in collaboration with state and federal land-management agencies, such as the US Forest Service and the US Department of the Interior Bureau of Land Management, as well as regional stakeholders groups, such as the nonprofit Mountain Studies Institute. All of the spatial data generated, including geology, watershed boundaries, rivers and lakes, wetlands, and mine sites, were originally produced on a CD-ROM as

Silverton, Colorado, one of the towns in the study area affected by 150 years of mining of silver and other minerals. Photo by Joseph Kerski.

part of a publication named Professional Paper 1651. The data are now available on the USDA website as a series of zipped ArcGIS coverages and shapefiles.

These fragile, high-altitude, mountain ecosystems, where much of the rugged terrain is over 10,000 feet, have been extensively mined for many decades. Today, the area is valued for its tourism potential and natural beauty. Present-day towns rely on water sources used extensively in mining operations to sustain them through the winter. To address the persistent problems of poor water quality in the area, prevent further damage to the environment, and safeguard health of the local residents, a GIS-based approach was critical to the initial assessment of the problem twenty years ago and remains critical today.

Summary and looking ahead

This chapter reviewed some of the broader aspects of working with public domain spatial data as part of a GIS-based project. We looked at some factors you should consider when choosing whether to use public domain data, and through a series of case studies, we provided an overview of some of the real-world projects that have successfully used public domain data.

In the following two chapters, we will examine some important new influences on public domain data: VGI and cloud computing.

References

Archy, Wanda. 2011. "Interactive Map Viewer Launched to Monitor Congo Forests." *Earthzine*, July 8.

Buxton, Herbert T., David A. Nimick, Paul von Guerard, Stanley E. Church, Ann G. Frazier, John R. Gray, Bruce R. Lipin, Sherman P. Marsh, Daniel F. Woodward, Briant A. Kimball, Susan E. Finger, Lee A. Ischinger, John C. Fordham, Martha S. Power, Christine M. Bunck, and John W. Jones. 1997. "A Science-Based Watershed Strategy to Support Effective Remediation of Abandoned Mine Lands." Paper presented at the Fourth International Conference on Acid Rock Drainage, Vancouver, B.C., Canada, May 30–June 6.

Croswell, Peter L. 2009. *The GIS Management Handbook*. Des Plaines, IL: The Urban and Regional Information Systems Association. 320.

Haklay, Muki, and Patrick Weber. 2010. "OpenStreetMap: User-Generated Street Maps." *IEEE Pervasive Computing* 7(4): 12–18.

Meissl, Stephan, and Gerhard Triebnig. 2010. "Land Monitoring Network Services Based on International Geospatial Standards: SOSI and geoland2/SDI Projects." *International Journal of Digital Earth* 3(Supplement 1): 70–84.

Pamuk, Ayse. 2006. *Mapping Global Cities: GIS Methods in Urban Analysis*. Redlands, CA: Esri Press. 208.

Peters, Dave. 2012. *Building a GIS: System Architecture Design Strategies for Managers*. Second edition. Redlands, CA: Esri Press. 384 p.

Tomlinson, Roger. 2003. *Thinking about GIS: GIS Planning for Managers*. Redlands, CA: Esri Press. 256.

8

THE DATA USER AS DATA PROVIDER

Introduction

In the early years of GIS, the data providers were invariably large federal mapping and science agencies and private companies, organizations such as Ordnance Survey, the US Geological Survey (USGS), and SPOT Image. During the 1990s, other organizations joined the ranks of spatial data providers, including the United Nations; federal agencies; provincial, state, and local governments; higher education departments and programs; nonprofit organizations; and an increasing number of private companies. While the data providers became increasingly diverse, they remained almost exclusively *organizations*—large-scale field mapping operations requiring considerable resources to invest in the technology for data creation and data delivery—not *individuals*. There was a clear distinction between data producers, who provided spatial information and services, and data users, who worked with and analyzed the data.

However, a convergence of technologies in the first decade of the 2000s altered the status quo. First, the advent of online data extended the scope of spatial analysis and spatial thinking, not only for the GIS data-using community but also the general public. Second, the removal of the Selective Availability error code in GPS signals in 2000 resulted in a boom in low-cost, easy-to-use GPS devices. Suddenly, ordinary citizens could measure any location on the planet within a few meters of its actual position using a low-cost GPS. Third, the explosion in the number of easy-to-use mobile devices made it possible to collect data without the previously onerous financial overhead of training and equipment. Fourth, as GPS became an embedded technology in many of these common mobile devices, such as cell or mobile phones, it became feasible to collect other data along with the positional latitude-longitude coordinates. Fifth, the emergence of web servers for data storage and hosting mapping services provided a repository and framework for the data to be uploaded and managed. Finally, greatly improved network data transmission rates broke down what may have been the last barrier—the ability to easily move the large files typically associated with spatial data from the source to the server.

Data users now have the capability to become data providers. The implications go beyond the greater number of organizations and online locations from which to obtain public domain data. *Data users as data providers* changes the entire dynamics of GIS. It affects every other issue we discuss in this book, including data quality, copyright, fees, scale, privacy, and standards. In this chapter, we will discuss the implications of the changes brought about by the thousands of individuals now participating in GIS as data providers.

What does data user as data producer mean?

In one sense, the notion of data users also producing data is nothing new. Even national mapping agencies occasionally dabbled with the idea of citizens contributing their local knowledge to the mapping process. The USGS, for example, ran Earth Science Corps during the 1990s and the National Map Corps program during the 2000s. Members of the public were asked to examine topographic maps of their area and suggest changes. Presumably, the maps with the most suggested changes would receive a higher priority for update. However such programs only allowed citizens to *suggest* changes, not *make* the changes themselves. If they did make changes, for example by annotating a topographic map or making a phone call to the relevant organization, those changes would have to be reviewed by professional cartographers who, in their role as gatekeepers, decided whether the new information was acceptable. If yes, they would then make the *actual* changes to the data. There was no guarantee that any suggested changes would be included in analog or digital mapping products. In practice, largely due to funding constraints, very few changes were ever made.

Today, citizens can do more than suggest changes. Their contributions can appear in data produced by traditional, and often conservative, data-gathering organizations. Perhaps more interestingly, citizens can increasingly bypass those organizations entirely and see their changes immediately reflected in the online databases they are updating.

This information gathered by ordinary citizens is sometimes referred to as *crowdsourcing*. In the world of GIS and spatial data, it is referred to as *volunteered geographic information* (VGI), or sometimes as *community mapping*, a term coined by Michael Goodchild of the University of California, Santa Barbara, in a paper he wrote for the first conference on the subject in December 2007 (Goodchild 2007). The broader context in which the spatial data-gathering occurs is typically referred to as *citizen science*, or *neogeography*. Jeff Howe coined the word crowdsourcing in his book *Crowdsourcing—Why the Power of the Crowd is Driving the Future of Business* as "the act of a company or institution taking a function once performed by employees and outsourcing to an undefined (and generally large) network of people in the form of an open call" (Howe 2008).

Despite the current focus on crowdsourcing, it is not a completely new innovation. Some high-profile initiatives have been around for more than a decade. Wikipedia, an online encyclopedia developed by anonymous, knowledgeable, and connected users, for a while was largely discounted as something that could not compete with the completeness and accuracy of collective reference works created by experts, such as a printed *Encyclopedia Britannica* or *World Book Encyclopedia*, or even a digital encyclopedia such as Microsoft *Encarta*. However, it could be argued that Wikipedia was the first crowdsourcing activity in which a large number of people participated and a large number of people could actually see and use its results. While all Wikipedia entries are supposedly reviewed by topic experts, the discerning user is wise to be critical of its content and find collaborative references elsewhere. That said, Wikipedia has emerged as a powerful source of information that has both breadth and depth. Much of the information it contains illustrates the premise that volunteers, when left to their own devices, can and will produce data that are both accurate and timely. As other consumers of the information are often experts in the topics they are researching, they are motivated to improve or correct entries that they deem inaccurate or incomplete. Over time, the content improves in both accuracy and scope.

Despite tremendous recent growth in VGI, one concern remains: VGI is still largely the preserve of those who have Internet access, a facility that is certainly improving but is by no means ubiquitous. VGI did exist before the advent of the Internet, but on a much smaller scale. One implication of the advancement of VGI and citizen science is the blurring of traditional boundaries between expert and nonexpert. Citizen science groups realized that GIS had enabled those who had access to the tools to influence and make decisions, and in many cases manipulate, GIS and spatial data to their own advantage. Once people realized they could indeed create and contribute data, they began to appreciate that they now had some influence and the opportunities for exerting that influence were almost limitless. This, in turn, has led to a multifold increase in data contributions each year. As place is such a local concept and one that has a great affinity with so many people, it was perhaps inevitable for maps and spatial data to emerge as one forum in which citizens throughout the world would readily participate.

Natural and human-made hazards and emergencies have been able to benefit almost immediately from crowdsourced data. The reasons for this are largely self-evident. Even today's most sophisticated earth-observing satellites may not be redirected over an affected area for several hours or days. Local conditions may prevent the use of existing digital data as the ground may be obscured or altered in some way due to fire-scarring, flooding, or landslides. There may be no access to the Internet due to power or IT infrastructure failures, or worse, because of injuries or fatalities. People living in the area are most familiar with the terrain and are increasingly able to report conditions to emergency services using smartphones with video, voice, pictures, and text.

Crowdsourcing is not unique to GIS. Book reviews on Amazon, seller's ratings on eBay, and tourists' reactions on TripAdvisor are all common examples of crowdsourced data. Mob4Hire Inc. and uTest Inc. provide a group of testers for companies who want to test software and mobile applications. In the scientific world, crowdsourcing has also become an embedded practice in many science organizations. For example, they include the amateur astronomer *clickworkers* who classify features on Mars images for NASA, those who map the locale and features of vernal (after rainstorms) pools of water, or those who locate invasive species for the US Fish and Wildlife Service.

Crowdsourced data also cover such spatial phenomena as crime, graffiti, traffic, and air quality. Crowdsourcing allows these phenomena to be reported more frequently than was previously possible by the responsible government agencies, given staff and budget constraints. Goodchild (2007) stated "it is easy to believe that the world is well mapped." However, as Estes and Mooneyhan (1994) noted, "in reality, world mapping has been in decline for several decades." The USGS no longer attempts to update its maps on a regular basis. A coauthor of this book worked at the USGS and watched his mapping program shrink from 525 employees to 110 over seventeen years. Many developing countries no longer sustain national mapping enterprises. This decline of mapping has many causes (Goodchild, Fu, and Rich 2007), including high cost, challenges in maintaining the funding of a multi-year effort, and the decreased need for field staff due to the availability of remotely sensed imagery. As a result of this decline, government agencies will look increasingly to crowdsourced data to supplement their data resources.

In the early 1990s, the Mapping Science Committee of the US National Research Council reported on the state of the spatial data infrastructure (NRC 1993), where the aggregate of agencies, technologies, people, and data constituted a nation's mapping enterprise was more of a patchwork than anything else. It recommended that national mapping agencies should no longer attempt to provide uniform coverage of the entire country, but instead should provide the standards and protocols under which numerous groups and individuals may create a composite coverage, varying in scale and currency depending on requirements. Many data-collecting agencies subsequently adopted this approach as their business model. VGI is a natural fit for this approach as it provides a body of people gathering data in response to the needs of local communities. The challenge will be to ensure this patchwork of different levels of accuracies, theme, and scales will emerge into a cohesive mapping entity. Or, perhaps a cohesive entity is not necessarily the goal. Does the entity necessarily need to be cohesive given the diverse set of requirements in the GIS community?

As citizen mapping initiatives and the accompanying web portals increase, the task of exhaustively listing all of them exceeds both the scope of this book and the ability of the authors to keep that list up to date! Instead, we will highlight a few of these initiatives to illustrate how and why they are transforming the field of GIS. We encourage you to search for other initiatives and portals and see for yourself what's available.

Citizen data collection initiatives

Some citizen data collection initiatives were taking place long before they had the ability to map and analyze their results. One such example is the National Audubon Bird Count, which has been taking place since 1900. The *Oxford English Dictionary* has also used amateurs for centuries to research the first published use of a word and expand the number of definitions for certain words. The challenge for many organizations was how best to communicate the results of their work. It's easy to see that it would be useful to map bird sightings but perhaps not word origins. But then again, why not? Who is to say that a map of word origins would not be of interest to a lexicographer or a linguistics researcher? James Cheshire, as a doctoral student at the Department of Geography and Centre for Advanced Spatial Analysis, University College London,

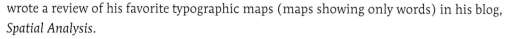

wrote a review of his favorite typographic maps (maps showing only words) in his blog, *Spatial Analysis*.

In the brave new world of VGI, it seems we are only limited in what we may achieve by the available mapping tools and people's time and imagination. However, with the advent of easy-to-use APIs and the availability of base data from Esri or other organizations, coupled with potentially thousands of people to do the work, perhaps even those are no longer limitations.

LandScope

LandScope America is an online resource for land protection and a collaborative effort between NatureServe and the National Geographic Society. Its vision is to publish hundreds of conservation maps at multiple scales, along with user-contributed geotagged articles, photographs, and videos. You can also contribute to the shape and content of the LandScope website itself. Its partners include DataBasin, another VGI-related tool developed by the Conservation Biology Institute, and the Conservation Registry, an online database developed by the Defenders of Wildlife to map conservation projects worldwide. Like many VGI-contributed sites, it is easier to contribute spatial data than to download data from the system. As is often the case, sometimes the best way to obtain the data is to contact the site coordinators.

ShakeMaps

ShakeMaps is one of the oldest examples of volunteered geographic information; it predates the current interest in VGI by at least a decade. It is a product of the collaboration between the USGS Earthquake Hazards Program and regional seismic network operators. The information it presents is different from the earthquake magnitude and epicenter (parameters which describe the earthquake source) data released after an earthquake in that a ShakeMap is a representation of the ground shaking produced by an earthquake. So, while an earthquake has one magnitude and one epicenter, it produces a range of ground-shaking events and levels at various sites. The location of those sites depends on distance from the earthquake, the rock and soil conditions at the sites, and variations in the propagation of seismic waves from the earthquake due to complexities in the structure of the earth's crust.

To provide more accurate and timely information on the location of the shake sites, the logical choice was to use a network of citizens to provide the density of data required. ShakeMaps provide near-real-time maps of ground motion and shaking intensity following significant earthquakes. These maps are used by federal, state, and local organizations, both public and private, for post-earthquake response and recovery, public and scientific information, as well as for preparedness exercises and disaster planning. Through the "Did you feel it?" project, citizens can report an undocumented earthquake using a simple form, with an optional street address that may be automatically geocoded. (Search for "did you feel it.")

FieldScope

FieldScope was created by the National Geographic Society and provides a web-based mapping, analysis, and collaboration tool to support geographic investigations. FieldScope does not cover just one area of the planet but includes many regions and many themes. Individual FieldScope sites have been developed using ArcGIS for Server, which provides both the data management and framework for user input. FieldScope focuses on students and education; here, students are engaged as citizen scientists to investigate real-world issues in both classroom and outdoor settings. FieldScope provides not only maps but mapping activities and a community where student fieldwork and data are integrated with that of their peers and professionals. This adds analysis and meaning to student investigations. Remaining local or regional in focus, FieldScope began with a study of the Chesapeake Bay watershed and now includes watershed dynamics, land use impacts on streams in national parks in the Washington, DC, region, and the biology of Biscayne National Park and Indiana Dunes National Seashore. Student work has been shown to be just as detailed and accurate as any other VGI, a point well illustrated by GLOBE (Global Learning and Observations to Benefit the Environment), a project that has been in place for nearly twenty years.

GLOBE

GLOBE is a worldwide, hands-on, school-based science and education program. It promotes and supports students, teachers, and scientists who collaborate on inquiry-based investigations of the environment and the working of the earth's systems. GLOBE's partners include NASA, the National Oceanic and Atmospheric Administration

(NOAA), and the National Science Foundation Earth System Science Projects to research the dynamics of earth's environment. The project began on Earth Day 1995 and has grown to include participants in 111 countries, 54,000 GLOBE-trained teachers, 23,000 schools, and over 1.5 million students. Over twenty-one million VGI measurements have been gathered since it began, including daily measurements of temperature and precipitation, data on water and soil chemistry, land cover, and phenology (how seasons and climate affect biological phenomena). Most of these measurements are available as point data that is easily modeled in a GIS environment, although some files require editing before they can be mapped and analyzed. GLOBE at Night is an annual citizen science campaign that records the brightness of the night sky, and the resulting map provides patterns on light pollution and population. Weather- and atmosphere-related projects like this are conducive to citizen science projects, and there are many examples, such as the Community Collaborative Rain, Hail, and Snow Network, whose goal is to have over 30,000 observers by the end of 2013.

BioBlitz

BioBlitz, as the name implies, is a rapid biological inventory of a specific site, which attempts to identify and record all species of living organisms at that site. Sites chosen are typically an urban park or nature reserve, and data collection is usually completed in a twenty-four- or forty-eight-hour period. The first BioBlitz was conducted in Washington, DC, in 1996 and the projects have since grown to include many small areas of the planet. They serve a dual purpose: to document biodiversity, as well as to help illustrate the importance of science and the gathering of scientific data. As BioBlitzes are localized, there is no one-stop-shop for downloading data, though a partial list of events can be found at the USGS website. Indeed, some of the results are not geocoded at all.

What's Invasive!

What's Invasive! is an example of a partnership between higher education and government, which involves UCLA, the National Park Service, and citizen input. The site seeks local input in the form of photographs and sightings of invasive plants and animals. A list of participating parks is provided, but citizens can add their own. Like an increasing number of VGI projects, What's Invasive! promotes citizen input via ordinary smart phones, and completely bypasses the need for handheld probes or even handheld GPS receivers. In common with an increasing number of VGI initiatives, What's Invasive! goes one step further than simply gathering citizen data; it offers citizens access to its open API, so they can contribute to the development of the project, as well.

Other examples of biology-related citizen science initiatives are the Colorado River Watch Network at the regional level and, at the national and global level, the Great Backyard Bird Count, Project BudBurst, World Water Monitoring Day, Project FeederWatch, and the Audubon Christmas Bird Count.

OpenStreetMap

Perhaps the best known VGI resource is OpenStreetMap (OSM). It is a free editable map of the world, available for all to use and built through contributions from volunteers, either working independently or collaborating with others at weekend mapping parties or on a Project of the Week. The OSM project, which began in the United Kingdom in 2004, was a response to some of the issues we discussed earlier in this book (such as the restrictive data access and pricing policies of Ordnance Survey) and quickly spread to Asia and North America. By November 2011, over 500,000 users were registered and more than 5,000 contributors were working each month, with the number doubling each year. OSM has become something of a social phenomenon, an opportunity for citizens to have a say and circumvent "the system." By late 2010, two books on OSM had been published; the subtitle of the latest, *Be Your Own Cartographer*, deftly captured the aims of the project (Bennett 2010). Fittingly, its publisher, Packt, donates some of the proceeds of the book to the OpenStreetMap project.

OpenStreetMap's content is far-reaching, despite its "street" name. It is not simply about mapping roads, but it also includes " . . . paths, buildings, heights, pylons, fences . . . AND. . . post boxes, pubs, airfields, canals, rock climbing routes, shipwrecks, lighthouses, ski runs, whitewater rapids, universities, toucan crossings, coffee shops, trees, fields, toilets, speed cameras, toll booths, recycling points, and a whole lot more." (Casey 2009). You can upload GPS traces to the website, view the history of edits to your area of interest, and contribute your own edits. As well as accepting data contributions, data can also be exported as GIS-ready OpenStreetMap XML data, as a Mapnik Image, an Osmarender Image, or as an embeddable HTML document. Although OSM is no longer unique (other emerging initiatives such as OpenNewYorkMap, OpenTreeMap, or Wikimapia.org have adopted the same approach), it remains truly a two-way street for uploading and downloading content.

KEEPING IN TOUCH WITH OPEN MAPS

Several contributors write for a blog and podcast called OpenGeoData, a resource to monitor open maps, open data, and OpenStreetMap.

Wikimapia

Wikimapia was launched in 2006, and like OpenStreetMap, is a citizen-generated map of the world. However, it adopted a gazetteer approach and describes features on the earth's surface, from cities to individual buildings. Each entry's geographic extents are defined by ranges of latitude and longitude and the accompanying website uses base-maps from other services for reference. Unlike OpenStreetMap, it is privately owned by two Russian entrepreneurs and does not make the data available under a Creative Commons license; all data contributed to the project are licensed to Wikimapia. According to the site's Terms of Service, all content uploaded by users becomes the property of Wikimapia and its future successors who retain all rights to future sublicensing and the potential transfer of licensing to other entities. At present, user submissions are available through the website for personal or educational noncommercial purposes, and may be used for noncommercial use through Wikimapia's own API, launched in 2009. The API is free for use at present, but the terms of agreement state fees may be imposed at any time.

By the end of 2011, over fifteen million places had been mapped in Wikimapia, and the site boasted nearly one million registered users. Some registered users join watch lists to monitor changes for their own area of interest, which helps to ensure any changes are legitimate. To prevent abuse of the site, users are classified into one of three levels, with level three reserved for a few hundred more experienced users. Although linear features added by users are limited to roads, railroads, ferries, and rivers, by far the most common added item is a point feature (a tag or a placemark location) to which the user may attach a photograph or a video. Another aspect of Wikimapia is that users can vote in favor of or against other users' contributions. This facility to promote or demote contributors through a ranking system has been in place in other crowdsourced sites and is now beginning to appear on mapping sites. Wikimapia also includes a user statistics and ranking board that automatically ranks users based on their contributions. Votes and user statistics, however, do not affect the user level of a contributor. Another useful feature of Wikimapia is that it is supports over 100 languages and the website interface has been translated into fifty-six languages. Tags do not have to be manually translated by the user; it is done automatically.

Citizen mapping networks

Even before the increased availability of easy-to-use GIS tools had an impact on VGI, citizen mapping networks had been expanding. This was due to a growing appreciation of GIS being much more than an expensive decision-making tool for the select few and having an almost unlimited range of applications.

MapAction is the only nongovernmental organization (NGO) that delivers vital support information through maps created from information gathered at the scene by a rapidly-deployed volunteer group of GIS professionals. These professionals gather a shared operational picture of the disaster so that humanitarian aid can be delivered to the right place as quickly and efficiently as possible. Since 2004, MapAction has helped in twenty emergencies, including the tsunamis, earthquakes, volcanoes, floods, and tropical storms. The volunteers participate in ten weekends of training each year in addition to their operational deployments. A cadre of full-time staff, part-time specialist officials, and a board of trustees support those in the field. MapAction strategic partners include the UN Office for the Coordination of Humanitarian Affairs (OCHA), the UN Institute for Training and Research (UNITAR) Operational Satellite Applications Programme, and the UK government Department for International Development (DFID). GIS is critical to the work of MapAction.

The GISCorps program of the Urban and Regional Information Systems Association (URISA), coordinates short-term volunteer GIS services to underprivileged communities around the world. It supports humanitarian relief, community development, local capacity building, health, and education. Begun in 2003, GISCorps has grown to hundreds of volunteers from over seventy-seven countries, and has deployed over 183 volunteers to sixty-one missions and thirty countries around the globe. The volunteers have contributed over 7,600 working hours towards those missions. One of the most notable efforts of GISCorps were the GIS and relief services provided to the citizens of New Orleans after Hurricane Katrina in 2005.

GeoMentor, launched in 2009 by National Geographic and Esri, was formed to connect thousands of GIS professionals with the educational community. Besides assisting educators in implementing GIS technology in schools, the project has the potential for connecting students, as interns, to local and regional governments, industry, and nonprofit organizations to help them collect spatial data for specific projects.

The mission of the Indigenous Mapping Network (IMN) is to "empower native communities by connecting them with the tools they need to protect, preserve, and enhance their way of life within their aboriginal territories." This is a good illustration of how embedded GIS has become in decision making and as a tool for enabling local communities to have more of a say in the matters affecting them most. IMN is a conduit for native individuals and groups to meet and build relationships, and to assist one another in accomplishing sovereignty goals. It endeavors to bridge the gap between traditional mapping practices and modern mapping technologies.

Many organizations are turning to VGI and GIS to showcase their content to the world through a universal language, maps. For example, National Geographic's Global Action Atlas spotlights local, cause-related projects from around the world. By making this available to a larger audience, many more individuals have the opportunity to take action by donating, volunteering, advocating, and sharing information. The atlas enhances and extends the mission of the National Geographic Society—to inspire people to care about the planet—by turning inspiration into action. However, caution is advised. National Geographic is both an organization (.org) and a business (.com) and some of the data on the site is not in the public domain. For example, the topographic basemap, National Geographic Topo!, was created originally from public domain USGS topographic maps, but with value-added features such as updated trails and shaded relief; this National Geographic product is sold and licensed and is not in the public domain. Furthermore, the data on many of these map-driven sites, such as the Global Action Atlas, are available only to browse and compile into online maps. Even if the data are available as downloadable KML files, the data may not be editable in a GIS environment.

At the beginning of the current decade, an increasing number of crowdsourcing initiatives began to use Ushahidi, a platform that enables collection, visualization, and interactive mapping of crowdsourced information on maps and in multiple channels, including the web, Short Message Service (SMS, or text messages from a cell phone), and Twitter. Built on top of the Ushahidi platform, Crowdmap allows people to collect information, aggregate it, and "see it on a map and timeline" without having to install Ushahidi.

An example of how these technologies may be integrated with Esri basemaps and GIS technology can be found in the data created for the 2011 floods in Australia. These web maps allowed citizens and decision makers to identify the location of the flooding,

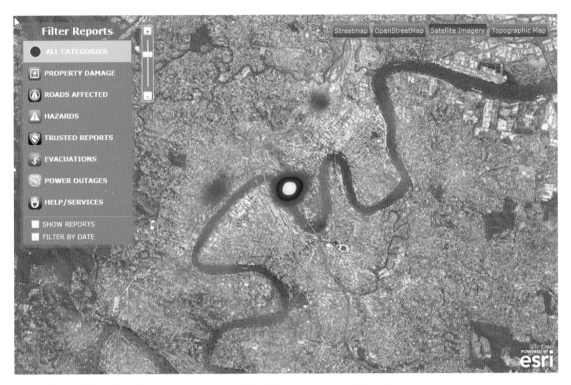

Map combining GIS cloud-based services, Esri basemaps, Ushahidi, and crowdsourced information to inform citizens and decision makers with current information about the 2011 floods in Brisbane, Australia. Image courtesy of NASA's Earth Observatory.

learn about the extent of property damage, roads affected, hazards, evacuation routes, and power outages, and make appropriate decisions based on that information.

Image archives

The availability of low-cost digital cameras and video recorders has resulted in an astronomical increase in the number of photographs and videos taken. With so many images and videos (and a compelling need to share them), various online image services were created by companies that could offer large amounts of data storage and fast data transfer rates. Although there are now many online photo archives to choose from, such as PicasaWeb and Panoramio, perhaps one of the best known is Flickr. Over three billion geotagged photographs had been uploaded to the site by 2008, with over five million posted each day.

Initially, although capturing details of specific locations, these images contained no spatial information. Tools have been developed to add location coordinates to the image or video metadata, provide a spatial reference, and geotag the image. This was achieved either manually with a text editor, with an Internet-based tool such as Geo-Imgr, or by using software such as Microsoft Pro Photo Tools. Automatic geotagging of images and videos has become increasingly common, as GPS-enabled cameras and smartphones proliferate. As a result, digital images and videos are now an invaluable spatial data resource. Much of the content on Flickr, PicasaWeb, Microsoft SkyDrive, and Google Docs could be considered a spatial data library.

Citizen science and GIS providers

An emerging trend combines citizen-gathered data with such authoritative sources as government agencies and private companies. For example, USGS collaborated with the US Forest Service, National Parks Service, Bureau of Land Management, and other agencies to create a wildfire map called GeoMac (Geospatial Multi-Agency Coordination Group) Wildland Fire Support back in the late 1990s. An innovation for its time, Geo-Mac offered near-real-time monitoring of wildfires via a web browser, providing home-owners, firefighters, and other non-GIS professionals with access to the information.

Another advance in the provision of public domain data has been the trend toward a shorter time lapse between a phenomenon occurring and when it is mapped. On May 18, 1980, David A. Johnston, a USGS volcanologist manning an observation post about 10 kilometers from the Mount St. Helens volcano, was immortalized with the transmission of his famous last message, "Vancouver! Vancouver! This is it!" before being swept away by the lateral blast created by the collapse of the volcano's north face. Nowadays, such real-time reporting is commonplace; spatial information is reported—daily, sometimes hourly, and sometimes in real time—not only on volcanic eruptions, but on wildfires, tsunamis, tornadoes, hurricanes, traffic congestion, weather, air and water quality, and many other phenomena. This is all possible because of web mapping services, the server farms where they run, Web 2.0 tools (a collection of Internet tools and technologies that allow people to actively contribute to websites), social media, and citizen data collection via mobile devices.

For example, Esri takes information from social media outlets like Twitter, Face-book, and Flickr, and translates the data into maps for first responders, providing a near-real-time spatial content resource that is not found anywhere else. Esri maintains an online wildfire map to provide continuously updated US wildfire information. The map also provides access to newsfeeds and precipitation information. Where does the citizen science come in? You can add your own information to the map, including links to websites, photos, videos, your own text, and other relevant information about rescue operations, evacuations, weather conditions, and current wildfire boundaries. To contribute, select a feature type under Shared Content, click the map, and provide a link to your content.

Esri has hosted a VGI-contributed web GIS application for just about every major disaster since 2010. For example, text messages, tweets (on Twitter), photographs, and videos were incorporated in mapping services set up for the Haiti earthquake, where people on the ground provided information about road blockages. This information was relayed to those distributing food and medical supplies. Esri also hosted an events map created during the 2011 political uprising in Egypt.

During and after the Gulf of Mexico oil spill of 2010, industry, represented by Esri, and four federal agencies—NOAA, the Environmental Protection Agency (EPA), US Coast Guard, and the Department of the Interior—integrated their data resources with VGI-provided data to host a map on the Geospatial Platform website that illustrated the trajectory of the spill, areas closed to fishing, wildlife data, research ships, and oil clean-up efforts. BP and the Coast Guard used the map to coordinate oil boom setup and oil-skimming vessel activity, while citizens used it to organize beach and wetland clean-up efforts. During its peak, NOAA provided public access to 325 data themes.

A wildfire in Boulder, Colorado, in September 2010 was a more localized disaster but, once again, citizen science through web GIS was heavily used. Homeowners used the wildfire maps to determine whether they should evacuate and when they could return to their property. Firefighters used the same maps to examine current fire perimeters, weather conditions, road access, and where best to concentrate their efforts. In the increasingly diverse mix of geo-enabled information sources, established news and information organizations are incorporating news reports from citizens, such as iReports on CNN. These web GIS response maps are sometimes difficult to locate, particularly after a disaster is over, because they are constructed largely for the non-GIS community to cover a specific event. They seldom contain downloadable layers or sophisticated analytical tools, but that could be changing; they may emerge as another valuable resource for GIS users.

Other long-established organizations are seeking input from VGI efforts. For example, the UN Food and Agriculture Organization (FAO) has conducted a remote-sensing-based global forest resources assessment. When the project needed a series of ground-based photographs to sample thousands of sites and verify the classification of its forest dataset, it turned to the Degree Confluence Project, a site hosting thousands of citizen-created photographs taken at intersections of full-degree latitude and longitude lines.

Combining VGI and government

A recent related trend could affect public domain spatial data users as much as, if not more than, any other issue discussed so far. An ever increasing number of government agencies are voluntarily combining resources and data to make their own operations more efficient. One example is a group operating in Victoria, British Columbia, Canada, that is integrating disparate geospatial datasets from 200 different organizations throughout the province—the provincial government, local governments, all major utilities, the property assessment authority, and other stakeholders—resulting in a multilayer geospatial fabric. The group decided to form a nonprofit organization, the Integrated Cadastral Information Society, and expanded their membership to include First Nations, the Royal Canadian Mounted Police, and the Department of National Defence. These last two organizations identified an increased need for geospatial data from a variety of sources because of their lead roles in securing and managing the 2010 Winter Olympic Games (Logan 2010). No federal authority ordered these organizations to collaborate; they recognized the requirement and decided to do it on their own. As budgets are constrained and more and more organizations see the value in geospatial data, examples of collaboration like this will become the norm, not the exception.

Citizen sensors

All around the world, a network of sensors—on satellites, on the ground, in the oceans, along fault lines and in the atmosphere—are busy measuring everything from vehicular traffic to air quality, from plate movement to ocean temperature. The number of global sensors increased so much after the year 2000 that Esri President Jack Dangermond began to refer to GIS as a "nervous system for the planet" (*ArcNews* 2001).

The Internet now provides access to many projects and resources for citizens to map where they have been. GPsies is just one example; anyone can upload their GPS-tracked trails (on foot, bike, by car, and so on) to an online mapping service. Other examples include Map My Tracks and Sanoodi. What if each of those contributing citizens was considered a sensor who moves about the landscape and collects data with handheld devices, or even without any device at all, just using their remarkable five senses? That network of thousands of artificial sensors pales in comparison to citizens as sensors in a human network comprising seven billion components. Each component represents an intelligent and mobile synthesizer and interpreter of information, documenting and recording in great detail virtually every aspect of our continually changing landscapes.

For example, commuter cyclists in New York City have been outfitted with a small sensor package that transmits the time, location, and air quality every day as part of the AIR (Area's Immediate Reading) project. With no funds to pay staff to monitor graffiti, citizens of Vienna have been provided with a new service for recording any racist graffiti they might come across. Incidents of graffiti are registered by submitting a geotagged photo to the Rassimus Streichen (Strike Racism) website. In other cities, trash containers and other facilities are monitored via a cell phone (or sensor attached to the facilities), which automatically transmits the toxicity, time, and location to a public website. Sensors like this are under development at various research institutions around the world including Urban Atmospheres and Common Sense—Mobile Sensing for Community Action. These efforts are not confined merely to a new way of collecting data. They are helping local authorities manage their resources better and rethink their data policies by contributing to online spatial data sources and extending the range of mapped phenomena. They also encourage people to analyze, share, and discuss the information they collect. Although an increasingly important resource for GIS users, be mindful of why and how these datasets were collected and managed, particularly if they are collected to influence environmental regulations and social policies.

READ AND RESPOND

Read the 2009 *New York Times* article "Online Maps: Everyman Offers New Directions." You may have to register (which is free) with this site to gain access.

What motivates Hintz, Kittle, and Ahmad to contribute thousands of map annotations? What is similar about the motivations of the three of them?

What is different about their motivations? How do their contributions impact the work of GIS users and that of ordinary citizens?

Data users as data advisors

Besides directly providing data for the GIS community, data users are more frequently able to influence data providers in indirect ways as well. One such example is through the advisory boards that make recommendations on regional, national, and international data policy. In the early days of national spatial data infrastructures, it was rare for data users to be included in the planning process, but nowadays it is a more frequent occurrence. In one such example, Karen Siderelis, geospatial information officer for the US Department of the Interior, asked a variety of citizen mapping groups to nominate key individuals to the National Geospatial Advisory Committee (NGAC). The NGAC is a federal advisory committee sponsored by the Interior Department under the Federal Advisory Committee Act, reporting to the chair of the Federal Geographic Data Committee (FGDC). The NGAC provides advice and recommendations related to management of federal and national geospatial programs, the development of the National Spatial Data Infrastructure, and the implementation of Office of Management and Budget Circular A-16 and Executive Order 12906. The Committee will review and comment upon geospatial policy and management issues and provide a forum to convey views of non-federal stakeholders in the geospatial community. One group that was asked to submit nominations was the IMN.

Volunteered geographic information data issues

Volunteered geographic information data provide a resource that government, private industry, and nonprofit organizations cannot match; a field staff of millions of people deployed around the world, ready to provide data in an instant. However, with the current and potential volume and sources of these data come concerns, as well.

Government agencies and private companies have collected a vast amount of digital data during the last thirty years and continue to do so at an even faster pace today. However, the amount of data collected through VGI will soon far exceed the traditional data resources. Will computing and network infrastructures, and even GIS analysts, be available to deal with the amount of available data? How can these disparate sources be indexed, searched, and used?

On the organizational side, how will VGI information be effectively integrated with the emergency management structure? So far, the information from citizens has been combined with services such as Esri's online ArcGIS basemaps, but they haven't truly been integrated into enterprise data repositories. It is largely up to the user to sort through the myriad of data types and information posted on Internet and social media sites. What happens if those sites are inaccessible? Facebook suffered an outage during September 2010 that lasted for several hours on two consecutive days and affected millions of users. For emergency services relying on contributions submitted via cell phones, what happens if the cell phone coverage is patchy or is disrupted by the very emergency or hazard citizens are trying to map?

Perhaps even more pressing are the concerns about data quality. The issues covered in chapter 2 are even more urgent when nonauthoritative sources are involved. How can ordinary citizens using ordinary equipment, such as GPS-enabled cell phones and low-end cameras, provide the level of quality required for GIS analysis? Comments made during the political events in Egypt in early 2011 raised questions about some YouTube videos and tweets being incorrectly geocoded. There are many reasons why social media can become incorrectly geocoded, such as poor GPS reception, intentional scrambling of location by the creator of the posting, national security concerns, or transmission errors. Given the sheer volume of information being generated and the lack of quality control, it is safe to assume most of these inaccurate readings will never be corrected. Along with the quality of the data, how can the authenticity and qualifications of the data provider be verified? What business practices will be established to ensure that data are checked and, if required for an emergency, can be verified quickly by knowledgeable people?

On the other hand, some argue that the type of data provided by VGI does not need to be survey-grade or mapping agency quality, nor can we afford to spend time and resources to verify it, as its value is chiefly in being in the right place at the right time. For example, VGI feeds provided rescue operators with critical information about roads destroyed after the earthquake in Haiti in January 2010 and helped direct rescue trucks via alternate routes or dispatch helicopters when roads were inaccessible. Some see citizens as authoritative sources because they are on the ground when a phenomenon is occurring and can create media to document what they are experiencing.

Not all data providers are wildly enthusiastic about VGI. Some in local government view VGI as a vulnerability in their day-to-day mission because allowing "just anyone" to edit spatial data exposes jurisdictions to potential liability. Another often-repeated comment from data providers is that "when everyone is responsible, no one is responsible." reflecting their fear of the "great unwashed using smartphones" making random edits to critical datasets. Significant legal issues such as privacy, intellectual property rights, data quality, and liability arise when VGI data are combined with data from an authoritative source. Many argue that sites such as Wikipedia have not, and will not, replace the need for peer-review.

This changing role of the citizen presents interesting research challenges:

What motivates an individual to volunteer?

What social factors could predict participation?

If a citizen could map anything, what would he or she choose to map?

What are the limits, if any, to the types of geographic data that can potentially be crowdsourced?

What factors determine the quality of crowdsourced information?

What mechanisms and protocols can be used to improve and assure quality?

Lastly, there are many spatial datasets that are not part of any established data infrastructure, but are routinely collected by individuals. One example is in precision agriculture. Farmers and ranchers have been using GPS-enabled tractor-mounted sensors to monitor their crops, yields, fertilizers, pesticide use, irrigation, and other inputs and outputs for over a decade. These data could be a rich resource for the entire agricultural science community, if they were made available online. However, if the data were placed on online, who would be responsible for removing personal information about individual farmers, ranchers, and their property? Even if the personal details were removed, could the spatial location information still be used to determine who contributed the dataset, and what they are doing, and using, on their land? Another example is personal health. Should a person's health records be tied to the location of their home so that emergency responders know the medical history of someone in need of immediate medical attention? Do the benefits of emergency medical staff access to this and similar types of information outweigh the potential invasion of privacy?

Open data policies

Even before citizen science emerged, many governments were under pressure to be more open with their constituents. Perhaps, in part due to the increased awareness of duplication and waste within government coupled with the improved communication facilitated by the Internet, it became easier for citizens to contact government officials. During 2010, GIS coordinators in several states watched with interest as legislation related to public data and online publishing began to appear, most notably, in the federal Public Online Information Act (POIA). The act would establish a Public Online Information Advisory Committee which would coordinate and encourage the government's efforts to make information from all three branches of government available on the Internet. It would also issue and update nonbinding guidelines on how the government should make public information available. It would direct the government to make public records available on the Internet at no charge, except as imposed by federal law before the act becomes law. The act requires the following:

- Public records are to be permanently available on the Internet.
- Current information technology capabilities are applied to the means by which records are made available and to the formats in which they are available.
- Public records are to be made accessible through programs and equipment that are readily available to the general public.
- Each federal agency to publish on the Internet must maintain a comprehensive, searchable, machine-readable list of all records it makes publicly available.

The Public Online Information Act delineates the roles of the director of the Office of Management and Budget, the administrator of the Office of Electronic Government, and the chief information officers of independent regulatory agencies. Exceptions to the Internet publication requirement are limited to those that are "narrow" and considered "case by case," if an agency requests an exception. The agency must demonstrate that there is clear and convincing evidence that the record should not be made available on the Internet, and on balance, the harm caused by disclosure significantly outweighs the public's interest in having the record available online. Finally, the Public Online Information Act directs the inspector general of each agency to conduct periodic reviews regarding agency compliance with Internet publication requirements and sets forth provisions regarding enforcement of public access by private individuals or organizations.

The Sunlight Foundation, which appears to be fostering some of these efforts, was founded in 2006 with a simple mission: To use cutting-edge technology and ideas to make government transparent and accountable. Its board of directors includes CraigsList founder Craig Newmark and advisory board member Jimmy Wales, co-founder of Wikipedia. The president of the National States Geographic Information Council, an organization dedicated to cooperation between state and federal agencies on the generation and provision of public domain spatial data, said, "This is another step toward openness and transparency. I have no idea whether the details are right or the legislation has a chance of passing. I'm glad Congress is talking about these issues. Public information is vital to our democracy and to our economy" (National States Geographic Information Council 2010).

Naming the legislation POIA appears to be a direct reference to the earlier generation's effort to pass FOIA (Freedom of Information Act) legislation. The group's name echoes a statement of crusading lawyer Louis Brandeis, who wrote in a *Harper's Weekly* article in 1913 that "Sunlight is said to be the best of disinfectants." He later became a US Supreme Court justice.

We have already discussed how the rise of VGI has spearheaded efforts advocating more GIS data become open to the public. However, the phenomenon has also led to citizens advocating that other data, which are spatial in nature but largely outside the realm of traditional GIS departments, become open as well. This includes public transit data. Advocates assert that citizens would benefit, streets would be safer, fuel would be conserved, and time and money saved by knowing even something as basic as when the next bus would arrive, no matter where you are located in a city, at any particular time. Authorative sources are already providing this type of transportation information, some of it in real-time. For example, Helsinki City Transport provides real-time data on the local bus network, Live Ship Maps tracks ocean-going vessel positions based on Automatic Identification System (AIS) data, and the Swiss Train Map, shows almost every train in the Swiss rail network in real-time.

Advocates for better access to transportation data recommend that the information be delivered in multiple forms, and most importantly for the convenience of citizens, delivered to their mobile phones such as Edinburgh's mobile bus tracker service. In many countries, government agencies lack the time, expertise, and funds to develop the applications to support such information delivery. Transit agencies are finding the easiest and least expensive alternative is to release data on routes, schedules, and real-time locations to software developers instead of guarding data like a proprietary secret. A comparison is often made with weather services. You can receive weather-related information in many formats—radio, television, web page, smartphones—as data from the National Weather Service are open and free. If that model could be applied to transit agencies, the public transport experience in many cities could be transformed.

In a related initiative, the National Public Transport Data Repository is the United Kingdom's largest transport dataset. Based on an annual snapshot of every public transport journey during one week in October, it provides information on where every bus stop, station, taxi rank, ferry terminal, and airport is located and how they are used. The data only come from authoritative sources, such as local public authorities, bus and train operators and, as of the time of writing, is not available in any GIS-friendly formats. It seems inevitable that public-transport-using citizens will want to contribute to this, augmenting the official version of transport usage with local insights.

One model for coordinating the activities of the many agencies involved in outsourcing work like this is the Contractor Portal from Dotted Eyes Ltd. The Contractor Portal automates and integrates the data sharing process among contractors, partners, and staff, allowing local government organizations to meet new imperatives to share data across expanding partner networks. It enables users to access an organization's data, and third-party data, such as those licensed under Collective Purchase Agreements, from Ordnance Survey, and other data suppliers. By providing data access on demand, users are able to make better, more locally relevant decisions. The creation of this portal reflects the ethos that access to data means better decision making.

Transparent government initiatives that include geospatial data as a fundamental component are continuing to emerge. The Chesapeake Bay Executive Council worked with Esri to build ChesapeakeStat, an online web mapping application that allows the public, Congress, restoration stakeholders, and project managers to follow the Chesapeake Bay Protection and Restoration program. An increasing number of these initiatives support data downloads and citizen input.

Data use policies

Some VGI data have proven so accurate that an interesting situation has arisen regarding data use policies. In particular, after OpenStreetMap data was shown to be quite accurate compared to Ordnance Survey data (Haklay 2008), the OpenStreetMap Foundation began to take steps to ensure that its data would not be used in commercial systems. One solution was to plant fake or *trap* streets, a method used decades ago by private map companies who wanted to ensure that no other company would copy its data. It would appear that we have come full circle. If OpenStreetMap were to find these fake streets used, say, as a base layer in an in-car navigation device, then OpenStreetMap could take appropriate action for a commercial provider infringing on its data. Wait a minute, you might say. Wasn't the intent of OpenStreetMap to provide a free and accurate dataset for anyone? It seems that geographic information generated by citizens has become more of a commercial product, altering established workflows and the traditional power bases in mapping.

VGI-driven sites are often a conglomeration of datasets (some contributed by volunteers and some contributed by organizations), data gathered at different scales, and data produced by different organizations. Determining whether any particular data are public domain is becoming more difficult to establish. For example, the LandScope site maintains data contributed by individuals, but some of it, such as ground-based photographs, are clearly copyrighted. Some organizations behind VGI sites are neither nonprofit nor not-for-profit corporations but rather a kind of social enterprise, as is the case of MapTogether. The resulting map-driven resources are more than just sites with maps, and more than just spatial information; they have become online communities. In the case of MapTogether, tools, training, and data are organized for community collaboration and knowledge sharing.

CloudMade was founded in 2007 to enable developers to build location-enabled applications and services, but one of the services they offer is OpenStreetMap data extracts. The data are licensed under the terms of the Creative Commons Attribution ShareAlike 2.0 license. You will recall from our discussion in chapter 1 that a ShareAlike license means if you alter or build upon these datasets, you may only distribute the resulting work under the same or similar license as the original data.

A decade ago, when the Internet was still relatively new, to access data you had to use the online resources of the organization that created the data. Nowadays, data are just as often served on a third party site by another organization, or on a site established by an individual who has obtained the data from elsewhere. For example, an individual at Nordpil obtained some data from the United Nations Population Division and in the process geocoded all the urban areas. The urban areas of the world, from 1950 projected to 2050, were made available online under a Creative Commons CC0 public domain license. However altruistic activities like this may appear, if you are considering using such a resource, always examine the license agreement and the data sources.

As VGI opens up new business opportunities, determining what are and what are not public domain data and applications will become even more complicated. CitySourced, for example, is a company that delivers an online solution for civic engagement using ArcGIS. Their tools allow smartphone users to take photographs of public safety and environmental quality issues and report those issues (along with a geocoded address) to the relevant local government. The site connects to a city's GIS layer using an ArcGIS 10 feature service from the Local Government template on the ArcGIS Resource Center. Just because the underlying data for applications like these come from somewhere else, don't assume they are in the public domain and you are free to use and reuse at will. Always check the source and licensing restrictions yourself.

Consider the implications of VGI for everything we have discussed in this book:

In terms of privacy, who is to regulate ordinary citizens collecting data on other ordinary citizens?

In terms of liability, what happens when people collect their own data, under circumstances where they could become injured while doing so, and which information potentially harms another person?

How can the data be accessed in the future?

How can these data be annotated with useful metadata?

How can VGI data be distinguished from government or industry-provided data if they all are placed into one large dataset?

How can VGI be integrated with traditionally collected information when it is from multiple sources, with multiple types of input, and with data collection taking place at different times?

In a position paper presented at a VGI conference, Kuhn (2007) argued for a new series of APIs to access and manipulate the VGI data.

Data quality

VGI is also sometimes referred to as asserted geographic information, as opposed to authoritative geographic information, because its content is asserted by its creator without citation, reference, or other authority (Goodchild 2007). Goodchild goes on to warn that, just as in the early days of the Internet, people didn't anticipate worms, viruses, and spam, but rather, altruistically believed in the goodness of others users. The same thing may lie ahead for VGI. At present, people are sharing data and enjoying the wealth of new information and possibilities for collaboration that such sharing brings. However, could erroneous spatial data be loaded on purpose to these data servers, with the intention of deliberately and maliciously influencing the results of any analysis and decision making based on the data? Will we see a new strain of computer viruses that incubate within data and on download maliciously modify the end user's computer? Such malevolent code would be very difficult to locate in a large vector or raster dataset. It makes us uncomfortable to think of such things, but who is to say the data we require will remain safe to use?

It is now so easy to create maps and spatial data, and to post those maps and datasets online. With the increasing volume of data comes an increase in the range of quality—some are of excellent quality, some of fair or poor quality, some of unknown quality, and some are just plain wrong. Established citizen science programs such as those operated by Audubon for counting birds and GLOBE for measuring precipitation, have developed a set of protocols and training programs to ensure quality, and to collect, synthesize, and redistribute the results. In addition, as social networking has grown up alongside VGI, it has become relatively common for the data user community to alert others about any omissions or mistakes in data. In an amusing but telling example of what may be more common in the future, Jessamyn West, a library activist and social media leader, found that for a period of time, her town of Randolph, Vermont, was placed in the middle of a lake by Google Maps. After her blogs, tweets, and Facebook posts failed to rectify the matter, her "My Town's In the Lake" song and video on YouTube attracted sufficient attention, even from *Time* magazine, to ensure that the mistake was corrected (NSGIC 2010). Publicity, good and bad, from a case like this raises awareness, and it is reassuring to learn that one data user can, eventually, correct an error in a dataset from a global data provider.

READ AND RESPOND

Read the 2009 *New York Times* article "Local governments offer data to software tinkerers." You may have to register (which is free) with this site to gain access.

Maps and GIS are becoming key components in many web and mobile applications based on government data that were previously not mapped or mapped in a way that was inaccessible to most citizens. Why do you suppose this is the case?

Next, investigate some of the sites mentioned, such as After School Special, CleanScores, or Trees Near You.

What sites do you find the most interesting, and why?

On which sites are spatial data available? Can you determine if the sites contain public domain spatial data?

According to the article, why couldn't software developers obtain a regularly updated feed of pedestrian and bicyclist injuries from the New York City Police Department?

Do you think the Police Department was justified in withholding the data?

What privacy concerns regarding this data request did the article leave untouched?

Consider both sides of this issue. There are plenty of opportunities for data users to submit suggestions and change requests to organizations producing or hosting the data. These organizations could be private companies, such as Esri via its ArcGIS Ideas portal. They could be government agencies, for example, Spain's Infraestructura de Datos Espaciales de España (National Spatial Data Infrastructure). Even governments are sponsoring mashup contests for citizens who wish to do something creative with maps and government datasets; the MashupAustralia Contest is just one example. Government data portals increasingly allow other types of input from data users. For example, Data.gov now includes a section where you can suggest a dataset for public release.

However, on the other hand, for those organizations doing the best they can to provide products and services with limited staff and budgets, widely-publicized complaints and highly visible errors in the data they provide can be an embarrassment, hand an advantage to competitors, or even jeopardize the support they have from within their own organization. A local government, for example, may not have time to incorporate citizen-corrected data, and devoting time to these corrections could mean that larger, more important projects are neglected. In these days of widespread spam and defamation claims, how do data providers know which complaints and suggestions to take seriously?

Contributing data to VGI sites often remains a one-way operation. Ideally, when you contribute data to an online resource, you should also get an immediate return on your investment of time and effort. For example, you should be able to quickly see your contributed data on the online map. You should be able to download updates immediately in a more reciprocal arrangement. Also ideally, you should be able to download data in a variety of spatial and tabular data formats, and you should be able to create and share derivative works from the data. Although these facilities are not always available, they are becoming more commonplace.

READ AND RESPOND

Start where it all began, the very first conference on VGI, held at the University of California, Santa Barbara, in December 2007. Examine the list of participants, scan the abstracts, and read the full papers.

What issues do you still find most relevant since this conference was held?

Summary and looking ahead

The GIS data user as a data provider is a relatively new phenomenon, but one that is already altering not only the experience of the public domain data user, but GIS in general. While the future may take further interesting twists and turns, one thing is certain, you need to be more diligent than ever before and think carefully about the data you find and incorporate into a GIS-based project. As further opportunities to contribute to spatial data repositories continue to emerge and the means to do so continue to become easier, the amount of spatial data available will increase exponentially. Some of the data will be useful; some will be useless. When people need information quickly during a crisis, there may not be time to provide metadata, much less verify the information. As with many resources on the Internet, once the crisis ends there may be no funding or project structure available to document or manage new data. Typically, the sites remain online because it takes less work to leave them up than remove them, but they are essentially abandoned. As VGI increases, so will the amount of data that is not documented. If data are undocumented, how can you decide whether or not to use those data in your project? You may be able to justify doing so for some projects, but for others, if the data are too suspect or the project requirements are too sensitive or demanding, it's not worth the risk.

Associated with VGI is a new information technology paradigm that is changing the face of traditional public domain data acquisition, access, and usage, and has propelled the volunteered geographic information revolution. This critical component is not, however, restricted to VGI and is becoming established as *the* new framework for providing access to online data and services. It is cloud computing. In the next chapter, we will examine the impact of cloud computing on public domain spatial data and, as with other topics covered in this book, invite you to reflect on the inherent advantages and limitations of the cloud environment as you use its resources to support your GIS-based decision making.

References

Bennett, Jonathan. 2010. *OpenStreetMap: Be Your Own Cartographer*. Birmingham, AL: Packt Publishing Ltd.

Casey, Michael J. 2009. "Citizen Mapping and Charting: How Crowdsourcing is Helping to Revolutionize Mapping and Charting." Conference proceedings, US Hydrographic Conference 2009, Norfolk, VA, May 11–14. http://www.thsoa.org/hy09/0512A_02.pdf.

Estes, John E. and Wayne Mooneyhan. 1994. "Of Maps and Myths." *Photogrammetric Engineering, and Remote Sensing* 60: 517–524.

"GIS Communities Are Poised to Take the Pulse of the Planet." 2001. *ArcNews* (fall). http://www.esri.com/news/arcnews/fall01articles/giscommunities.html.

Goodchild, Michael F. 2007. "Citizens as Sensors: The World of Volunteered Geography." *GeoJournal* 69 (4): 211–221. http://ncgia.ucsb.edu/projects/vgi/docs/position/Goodchild_VGI2007.pdf.

Goodchild, Michael F., Pinde Fu, and Paul Rich. 2007. "Sharing Geographic Information: An Assessment of the Geospatial OneStop." *Annals of the Association of American Geographers* 97(2): 249–265.

Haklay, Muki. 2008. "How Good is Volunteered Geographical Information? A Comparative Study of OpenStreetMap and Ordnance Survey Datasets." *Environment and Planning B: Planning and Design* 37(4): 682–703.

Howe, Jeff. 2008, 2009. *Crowdsourcing: Why the Power of the Crowd is Driving the Future of Business*. New York: Random House Digital, 311.

Kuhn, Werner. 2007. "Volunteered Geographic Information and GIScience." Position paper for the NCGIA and Vespucci Workshop on Volunteered Geographic Information, Santa Barbara, CA, December 13–14. http://www.ncgia.ucsb.edu/projects/vgi/docs/position/Kuhn_paper.pdf.

Logan, Barry. 2010. "The ICIS: Achieving Collaborative (Geospatial) Success." *Directions Magazine*. April 5. http://www.directionsmag.com/articles/the-icis-achieving-collaborative-geospatial-success/122385.

National Research Council. 1993. *Toward a Coordinated Spatial Data Infrastructure for the Nation*. Washington, DC: National Academies Press.

National States Geographic Information Council. 2010. "What To Do When 'Public' Data Is So Publicly Wrong?" (June 11). http://news.nsgic.org/2010/06/what-to-do-when-public-data-is-so.html?utm_source=feedburner&utm_medium=feed&utm_campaign=Feed%3A+TheNsgicBlog+%28NSGIC+News%29.

9

PUBLIC DOMAIN DATA ON THE CLOUD

Introduction

As you have learned during the course of this book, despite great advances in hardware, software, and the variety and availability of spatial data, the business and workflow model for GIS remained relatively unchanged throughout the 1990s and 2000s. During the early decades of GIS, public domain spatial data were initially available on various forms of physical media. This included 9-track magnetic tape, zip drives, CD-ROMs, and floppy disks. Those working in the early days of GIS may shudder to recall these media which were, at times, difficult to use and share, and often failed. GIS data became increasingly available via the Internet during the 1990s, liberating users from relying on physical media for data delivery.

As the Internet became more ubiquitous, small vector datasets began to appear online; improved data transfer rates meant that larger raster datasets soon followed. Providing spatial data on the Internet meant faster and more efficient access to data. Although these changes opened up new applications for GIS and expanded the number of users working in traditional GIS areas, they did not usher in a new business model for spatial data. As the agencies simply provided the same data online that they had formerly shipped to users via physical media, the only thing that changed was the method of delivery. Instead of loading data from a cartridge or disk, the data were downloaded from the web. For those working with the data, these were welcome changes. The data were processed and analyzed no differently but were easier to access. For data providers, too, this was also a welcome change, as they no longer had to process user requests for data on physical media. Yet, the predominant operational procedures remained unchanged; GIS users wanted the data stored locally so they could access the data when they chose. However, within the last couple of years a new technology has emerged that will have a profound impact on GIS and alter forever the thirty-year paradigm of how spatial data have been delivered, accessed, and utilized. That technology is cloud computing.

In this chapter, we will define cloud computing and examine why and how it fundamentally alters the way public domain spatial data are accessed and used. We will spend some time discussing a particular application of cloud computing—using ArcGIS Online—that illustrates the impact this new technology and data access model has on the GIS user community. In the previous chapter, you reflected on the advent of volunteered geographic information (VGI). Contributing to and downloading VGI information is largely made possible because of cloud computing. Probably more important is the ability to easily combine or *mashup* online maps using online web applications. You can stream data from the cloud into your desktop GIS application. You can download data from the servers that support the cloud, as you have done earlier in this book. You can also make your own data available to the cloud as you work with others to solve spatial problems.

What is cloud computing?

The term cloud computing refers to a network of servers; when the first diagrams appeared illustrating a configuration of servers, they often used clouds to represent computers that were not physically in front of a user. Rather, cloud-based computers or servers were "up there" or "out there," hosted in a different part of the user's organization, or even more commonly, by another organization altogether. They were accessed, but generally never seen, by the end-user.

According to the National Institute of Standards and Technology (NIST) (Mell and Grance 2011), cloud computing is a "model for enabling convenient, on-demand network access to a shared pool of configurable computing resources that can be rapidly provisioned and released with minimal management effort or service provider interaction." NIST includes networks, servers, storage, applications, and services in the components of a cloud, along with five characteristics, three service models, and four deployment models. The essential characteristics include the following:

- On-demand self-service that does not require human interaction with each service's provider
- Capabilities accessed through standard mechanisms over the network that promote use by a variety of clients, such as mobile phones, PDAs, and laptop computers
- Resources (such as storage, bandwidth, and memory) that are pooled to serve multiple users, assigned and reassigned according to demand

You, as the end-user, have no control or knowledge of the exact location of the resources but can use the system for specific requirements. Along the same lines, its capabilities are rapidly elastic and, to the user, may appear unlimited. Cloud systems control and optimize resource use by metering services, which allows for transparency for both the provider and consumer of the service.

Throughout its nearly fifty-year history, GIS has undergone some fundamental shifts in implementation. The evolution from mainframe computers to minicomputers from the 1970s to the 1980s, from minicomputers to desktop computers from the 1980s to the 1990s, and now the migration of GIS from the desktop to remote servers from the 2000s to the 2010s. Cloud computing is the next stage in that evolution.

For GIS practitioners, the cloud should not be considered as simply a big data repository "out there" nor simply the transference of data and tools from the desktop computer to computers across town or halfway around the globe. Rather, it represents a dynamically scalable architecture that is based on user needs. It is the new implementation and operating model for GIS—a model based on virtual machines that enables you to access data and run software remotely, access services and clients, and host content. Cloud computing has the potential to have a greater impact on GIS than any of the previous technological shifts for several main reasons.

Advantages to cloud computing

One big advantage of cloud computing is that it promotes availability of spatial data, which is central to the effective use of GIS. For GIS data users, the cloud is already becoming an invaluable source of spatial data; an increasing number of spatial data archives are offered—not just for download, but also as map services. When data are made available as map services, they may be streamed as tiles directly to end-users. For example, World Imagery and US Geological Survey (USGS) topographic maps are now hosted by ArcGIS Online, a cloud-based library of data and tools that we will discuss in more detail later on. In another example, the aerial photographs from the National Agricultural Imagery Program (NAIP) discussed in chapter 3 are now available online.

However, even more important as a driver for cloud computing than storing large amounts of data on the cloud is the requirement to model complex processes. The complexity of modeling, especially real-world phenomena such as climate, hydrologic, and ocean processes, requires high-capacity processing facilities. The cloud offers this with the additional advantage of providing access to these facilities only when needed. The significantly reduced processing time for analysis involving a number of what-if scenarios, many inputs and outputs, large datasets, and applications that use 3D modeling, visualization, and simulations, may be a compelling reason to consider performing that analysis in a cloud environment. An example is the European Space Agency's Grid Processing on Demand (G-POD) cloud project that processes large amounts of earth observation data.

ACCESS NAIP IMAGERY

For those working with ArcGIS, to access the NAIP imagery use the Add Data tool in ArcMap. Select GIS Servers, select Add ArcGIS Server, select User GIS Services on the Add Server pop-up window, accept the default of Internet, and enter the web address: http://www.usda.gov/. Once you have added the data, the service will be available, pointing you to the data directory that includes NAIP imagery, Conus, and Maps. Navigate to the NAIP directory and select your state of interest. Some states, such as Colorado, have posted both the color infrared and natural color imagery acquired in 2009.

NAIP imagery is an example of a traditional data source that is now available as a cloud service. It is still available from NAIP imagery webpage of the USDA, but is also now available as a service. For anyone wishing to access the data, the only difference is how the data are used. If you prefer to have a copy of the data to analyze locally, then download the data from the USDA website. If you wish to simply use the data as a basemap image as background context for your other data, then stream the data into your application from the NAIP data service. The advantage of the first method is you have access to the data as a locally stored file that can be used offline. The disadvantage is that you may have to download many files depending on your study area, which takes time and storage space. The advantage of the data service is that you do not need to download and store the data; you just use them. The disadvantage is you have to be online to access the imagery data.

At present, cloud computing has been implemented as three main service models, although bear in mind that cloud computing is rapidly evolving and other models are emerging. The first model is *software as a service* (SaaS). Also known as software on demand, SaaS allows end-users to use applications running on a remote server, most typically through a *thin-client* interface such as a web browser. The user does not access, control, or manage the supporting network, servers, operating systems, applications, or storage. GIS vendors may license applications on a subscription model, so users pay as they go. According to Philip O' Doherty, CEO of eSpatial (O'Doherty 2010), it is a fundamental change for the GIS community, because users and organizations can think about paying for services to achieve their goals, rather than buying software and maintenance contracts. To keep up with these new demands, GIS vendors may need to start thinking like service providers, not software companies.

The second cloud model is *platform as a service* (PaaS). This allows service users to upload applications they developed to the cloud infrastructure. The user may have some control over the application, and possibly the environment that hosts the application, but not the cloud's underlying network, operating system, or servers. The third model is *infrastructure as a service* (IaaS), where the service user does not control the hardware but does control and run software. In doing so, the service user influences operating systems, storage, and applications, with limited control over some aspects of networking, such as firewalls.

Examples of these three models in use today have proliferated. The SaaS model has been adopted by Salesforce.com, a company specializing in providing applications and data for sales representatives, executives, and managers. The company's telephone number—1–800-NO-SOFTWARE—confirms that the entire business model is based on no local software installations; all of the data and applications are stored in a cloud environment.

Esri provides a GIS-based example of SaaS, as it now offers GIS software capabilities in the cloud such as routing and geocoding services that may involve thousands of addresses and large amounts of linear vector data. Providing processing options like these as a cloud service enables organizations to geocode and map addresses or run routing models to generate possible routing scenarios much more quickly than was previously possible. It also allows organizations or businesses to submit these tasks to the cloud environment as a series of jobs that do not use any local computing resources and may run concurrently with other work undertaken locally. Another example of SaaS in GIS is Esri's Business Analyst Online, which allows users to combine spatial analysis with extensive demographic, consumer, and business data to make decisions such as optimum store locations or where to market a new product.

PaaS are database tools and process management tools that provide a middleware service for developers. In GIS, a common example is a set of APIs used by developers to create custom maps and queries. APIs represent a radical change in GIS development. By late 2010, Esri was supporting platforms, including the web (JavaScript, Silverlight, and Flex), desktop (WPF and Flex using Air), and mobile (JavaScript, Silverlight for Windows Phone, and iOS). What this means for GIS is that APIs make commercial-grade GIS technology accessible to users who are not specialists in GIS. Platforms and devices can take advantage of these APIs to support geo-enabled applications in the field, such as ArcGIS for Windows Phone.

Esri provides an example of the third cloud model, IaaS, having developed partnerships with cloud vendors, such as Amazon, and cloud-enabled supercomputing sites, such as Rocky Mountain Supercomputing Centers in Montana (*Inside HPC* 2010), to provide IaaS services. IaaS are also referred to as *off-premises solutions*, as no processing

READ AND RESPOND

Read the article in the online publication *Directions Magazine* titled "Every Cloud Has a Silver Lining; Only 'GIS as a Service' has Potential to Meet Growing Demand for Geospatial Technology." In it, O'Doherty (2010) proposed the term *GaaS* or GIS as a Service. He went so far as to say, "Someday, all software will run online and clunky desktop applications will be a distant memory."

Why does O'Doherty believe that GIS on the cloud has the ability to push GIS beyond what he calls "its existing niche markets?"

Do you agree with him?

What other forces are acting on society and in technology to affect the speed and potential of GIS to truly reach the masses through the cloud?

takes place at an organization's premises or facility, and as *on-demand services*, because they only run when the organization requests them, rather than as dedicated services.

Cloud deployment models

At present there are three deployment models of cloud computing: private, community, and public. Private clouds are operated for one organization only and are not available for general access. A city or federal government agency, for example, could operate a private cloud for the GIS users of its organization. The infrastructure may or may not be on the premises of the organization itself. For example, the US Environmental Protection Agency's cloud environment hosts its air and water environmental data for in-house analysis. This is quite separate from the public-facing data libraries that they contribute to via their main website.

A community cloud may be shared by several organizations. A community cloud, as the name implies, supports a community of users with shared concerns, missions, security clearances, compliances, policies, scales, or levels of analyses. A community cloud for climate studies, for example, may be set up and shared among analysts in oceanic and atmospheric sciences in government, academia, and international non-profit organizations. Community clouds may be accessible to GIS users who are a part of that specific research, development, or policy community. A public cloud is probably of most interest to GIS users. As the name suggests, it is accessible by the general public, or very large communities, and is typically owned by an organization that sells or offers cloud services.

Finally, any of these cloud models may be combined into a hybrid cloud, which allows organizations to adopt a mixture of in-house and outsourced computing and networking resources. For many adopters, this provides the most flexibility and allows them to take advantage of cloud technology, while at the same time support legacy data centers and systems.

Clouds offer many advantages to GIS users, not least of which is a greatly improved user experience. As the cloud is based on the collective computing resources of hundreds of servers, the end-user experience is faster and more dynamic than it could ever be on a single desktop or even a group of linked desktop computers. Cloud computing allows GIS users to access resources through easy-to-use tools via a web browser interface. Cloud architecture is modular; individual modules can communicate with each other yet remain independent, which means they can be taken offline for maintenance or repair without interrupting the services provided.

Cloud computing also has the advantage of *statelessness*; each request to the server is treated as an independent transaction and is unrelated to, and independent of, a previous request. This simplifies the configuration of the servers, as there is no requirement to dynamically allocate storage to deal with *conversations* in progress. If a client connection fails in the middle of a transaction, no part of the system needs to be responsible for restoring the present state of the server. Finally, cloud computing offers *semantic interoperability*, which enables the receiving system to correctly interpret information accepted from the transmitting system; the two systems will always interpret information in the same way.

It is too early to measure the full impact that cloud computing will have on GIS and public domain spatial data. However, if developments over the past couple of years are anything to go by, the potential impact is large and pervasive, affecting just about every workflow in GIS in the way most datasets are hosted, accessed, and used. For those organizations that found the entry to GIS too expensive for the kind of computing power they needed, the cloud offers great potential for them. It provides the opportunity to do what they have always wanted to do with spatial data storage and analysis without the onerous expense of purchasing hardware and software. The cloud offers a pay-as-you-go model that allows prospective cloud customers to try before they buy and allows existing customers the ability to pay in advance and take advantage of volume discounts. As such, the cloud is democratizing GIS; it enables access to supercomputing for those who need it but were previously unable to do so.

On a much broader scale, cloud computing has the potential to do what many in GIS have long discussed: elevate GIS from its present role as a modestly successful, but often under-appreciated, business component to an essential component at the heart of *all* business applications.

The Amazon cloud

Cloud platforms can run applications, store spatial and other data, and provide other services for both data users and developers. Cloud technology is now delivering core spatial datasets that were previously only available for download as static tables or maps. While several data providers offer public cloud platforms, Amazon has emerged as the leader in this area with the development of Amazon Web Services (AWS) and its vast network of servers.

To run applications, AWS essentially creates virtual machines through its Elastic Compute Cloud (EC2) service. GIS developers can request a virtual machine or an instance install software and just start using it. To store data, AWS can use a simple storage service, a relational database service, or other services. During 2010, public domain spatial data began appearing on AWS when the TIGER/Line shapefiles for the entire United States, approximately 125 gigabytes of data, became available.

Obviously, datasets of that size are not going to be downloaded by most GIS users. However, new technologies are developing, such as Amazon's Elastic MapReduce that enables data users to easily and cost-effectively "process vast amounts of data," according to Amazon. This particular utility uses a hosted Apache Hadoop framework, which supports data-intensive applications under a free license, running on the web-scale infrastructure of Amazon EC2 and Amazon Simple Storage Service (Amazon S3). It can be applied to spatial data just as easily as to other data and, since GIS data are often bulky, this has the potential to make available previously inaccessible large datasets. AWS is providing access to not just spatial data but a rapidly expanding resource of data for other sciences and policy makers.

In 2010, Esri joined the growing community of AWS Independent Software Vendors building services and solutions in the cloud computing environment. It made sense for Esri to work with cloud-based technologies because its software was already available on all other major platforms, including servers, desktop, and mobile devices. The aim is to provide a standards-based platform for spatial analysis, data management, and mapping, so that data users can take advantage of the cloud's security, scalability, and ease of deployment. A big challenge for smaller organizations wishing to migrate to a server-based deployment was finding the staff and resources to install and maintain the software. The Amazon EC2 service environment means that they won't have to do either because the installation is preconfigured and maintained on the cloud. However, the services can be administered and controlled by the local organization, just as if the server were located somewhere within their own organization.

Esri's initial use for the cloud environment was in four main areas. First, users could run ArcGIS for Server on AWS; when ArcGIS 10 was released, Esri made ArcGIS for Server available on the EC2 cloud as another option for deployment. This provided advantages of easier and faster deployment, lower cost for organizations, simpler development and testing, and better performance for the end-user. Second, users could access and contribute to ArcGIS Online, a service that we will return to later in this chapter. Third, ArcLogistics was ported to the cloud for optimizing route processing and analysis. Fourth, Esri products based on an online model were ported to the cloud. The first was Business Analyst Online, for analyzing demographic, consumer, and business data. In 2011 Community Analyst, for spatial decision making based on consumer and demographic data, was added. As an example of the enormous advantages of the cloud, Community Analyst included over 6,000 variables when launched.

During June 2010, the City of Novi, Michigan, (population 53,000) teamed with Esri partner Geographic Information Services, Inc. (GISi) to develop one of the first ArcGIS 10 Amazon Cloud-hosted applications. An Internet mapping portal for the city provides geospatial information, not only for its GIS users but also for its residential and business communities. The system was not created from scratch but rather was created by migrating the city's old ArcIMS (Internet Map Service) platform to ArcGIS for Server using the ArcGIS API for Microsoft Silverlight version 2.0. The result, hosted on Amazon EC2 in a secure environment, allowed the city to provide access to fifty map layers representing ten themes of information, including community and economic development, recreation opportunities, land ownership interests, ordinances, voting precincts, and public safety incidents. One of the goals in the migration was to promote the growth of existing businesses and to attract new investment within the community. Cloud-based GIS was seen as not just a service for GIS users, but a marketing tool. Such examples will no doubt become commonplace in the years ahead.

How does the public domain data user take advantage of Amazon cloud data? Public domain datasets can be viewed on the AWS page, which includes the US Census data mentioned earlier, and it seems only a matter of time before more spatial datasets will be available in the future. These public datasets are hosted on Amazon EC2 for free as Amazon Elastic Block Store (Amazon EBS) snapshots. Amazon EC2 customers can access this data by creating their own personal Amazon EBS volumes and using the public dataset snapshots as a starting point. They can then access, modify, and perform analysis on these volumes directly using Amazon EC2 instances and pay only for the compute and storage resources they use.

How cloud computing is changing public domain data and GIS

What do all these developments mean for the GIS user? As we have already mentioned, it means that an increasing amount of data will be accessible via faster channels and easier-to-use interfaces. It also means that, for ArcGIS users in particular, data will be more easily obtained than ever, as Esri's ArcGIS can be run on cloud platforms, such as Amazon EC2. However, as we discussed in the last chapter, today's data users are also data providers, so sharing data will perhaps be just as important as accessing data. The AWS platform invites data users to consider sharing a public domain dataset, as well, so as the service develops and becomes more widely used, it should develop into a world-leading spatial data resource.

Data

Cloud computing is changing spatial data in ways that the GIS user community is only beginning to understand and adopt. First of all, the cloud has blurred the lines between consumer map data and data used by the GIS community. For several years after the introduction of street and image maps for consumer applications such as MapQuest, Google Maps, and Yahoo! Maps, these datasets were typically used as base images, providing context for other datasets viewed online. As a data resource, they were unusable in GIS because their georeferencing engine was different from that used in GIS applications. Additionally, these data sources were available online, while the bulk of GIS use was still on desktop computers with data accessed locally. Desktop and online data could not be easily combined and data were invariably in different projections or could not be displayed in the same data frame. In the short time span of a year, all that has changed. Cloud-based services now routinely host GIS-ready georeferenced basemaps streamed from cloud-based servers to the desktop and integrated with other spatial data through GIS applications while, at the same time, remaining available for a much larger community of data users.

Second, cloud computing is built on a foundation of hardware and networks. Large arrays of servers with high-speed connections provide quick and easy access to data, bringing major computer companies and the GIS community together in strategic partnerships. Microsoft and Google are the obvious examples. Some see the competition for the consumer market between some of these companies as beneficial for GIS users. The competition has produced new and improved GIS tools and data viewers and has enabled consumers to acquire basemap data. These companies can leverage their existing server arrays to host data and software services running on those arrays. For example, Yahoo! uses an open-source cloud infrastructure to serve some of its map data.

Another important benefit of cloud computing for GIS users is, as we mentioned previously, that the cloud is already "home base" to nearly all volunteered geographic information, with VGI being served to data users on an ever-increasing basis. Some crowdsourced data are now being made available by private companies as data services. For example, MapQuest's Map Builder application includes OpenStreetMap data.

Finally, despite huge increases of the amount of data available from traditional data providers and users, these will only be a trickle compared to the explosion of data that we will see in this coming decade. Indeed, technology observers like Higginbotham (2010) say that the true era of "big data" will arrive as sensors become embedded in common objects such as roads, airplanes, and home appliances, each of which have the potential to generate terabytes of information daily. All of those data have to be stored, managed, and made available somewhere; cloud computing provides the infrastructure and environment that makes that possible.

Data delivery

Core spatial data are now delivered in new ways due to the advent of cloud technology. For example, when stream flow and water quality information from USGS stream gauging stations was first available online, the information was delivered via thousands of static web pages. That required a great deal of manual copying, pasting, and editing by the end-user to transform the data into something that could be mapped and analyzed within a GIS environment. StreamStats, developed by the USGS and Esri, solved many of these problems by developing a map-based web application that provides information on up to 165 basin characteristics (including stream slope, mean annual precipitation, and drainage basin) and up to 500 stream flow characteristics (including peak flow, annual and monthly means and medians). In addition to making the data available in a much more easy-to-use manner, StreamStats also provides a more intelligent solution: If a user selects a location where no stream gauge exists, a cloud-based GIS procedure will measure the basin characteristics and estimate stream flow at that location.

Private companies are also taking advantage of cloud technology to serve data. Satellite image provider DigitalGlobe, Inc., hosts a web feature service, a map service, and a map tiling service to provide access to their vast image library. With the ArcGIS extension called Image Connect, ArcGIS users can stream and download satellite images directly from DigitalGlobe into their desktop GIS application.

CONNECT TO A CLOUD-BASED SERVICE

It is easy to connect to cloud-based services for spatial data. For example, if those services are running ArcGIS for Server, you can connect to them from ArcMap. First, double-click GIS Servers in the catalog tree and then Add ArcGIS Server. Select Use GIS Services and select the connection type and the server to which you want to connect. If you are connecting to a server on the local area network, select Local and type the name of the server you want to connect to. While your GIS server may be configured over several machines, the server machine you should specify is the one that's running the server object manager. If you are connecting to a server through the Internet, select Internet and type the URL of the server you want to connect to. URLs follow this pattern: http://<server name>/<instance name>/services. For example, http://myServer1/ArcGIS/services. Finally, select Finish to connect to the server. The instance will now appear in the catalog tree, allowing you to add data from the server as you would from any other source.

Software

New types of software can be made available because of the cloud infrastructure services. For example, in 2009, Esri introduced MapIt, a software solution providing geocoding and mapping services. With MapIt, analysts could integrate geographic data with information stored using Microsoft SQL Server data management software. MapIt was also offered as a cloud service that allows organizations to access their data stored on the cloud and use their existing systems to visualize and analyze data. These existing systems include SQL Server, Microsoft Office SharePoint Server, and the Silverlight web browser plug-in.

API development platforms

The three most common APIs in use for creating web-based maps are Javascript, Flex, and Silverlight. Javascript and Flex have been out for awhile having previously been used in the development of such things as animations and videos on the web and increasingly now used for GIS. Esri worked with Microsoft to take advantage of Microsoft's

SQL tools and Microsoft's SharePoint Server through the Silverlight API. With these APIs, you can add content to an online map via existing services, customize an online map by adding your own tools and buttons, build stand-alone applications, add web content such as GeoRSS feeds, and connect to GIS services. Development with the APIs is free; deployment of the resulting applications and tools in the cloud is free under certain conditions. Is it realistic to expect that most data users will start using the APIs to develop mapping applications for their data online? When the Internet was new, people at first browsed web pages and consumed its content; a very one-sided process as users had little opportunity to interact with the sites they visited. It wasn't long, however, before individuals and organizations were no longer satisfied with consuming and, increasingly, they wanted to create their own web pages. It was, and remains, easy to view the HTML source code of any web page. The behind-the-scenes programming was transparent and could easily be adapted by copying and pasting code from existing pages into new pages, then modifying it to suit. Even people who had previously not considered themselves programmers were able to develop HTML code.

This same is now true for web mapping APIs. Sample ArcGIS code is available online for users to get started, so once again it is a matter of using simple copying-pasting-editing techniques to reproduce the desired results. The API interfaces make any existing application transparent as well, enabling a growing number of data users to provide access to their own data. While most data-users-as-providers won't go on to become an Ordnance Survey or a National Oceanic and Atmospheric Administration, they are in a sense performing a similar role as they make maps with existing data and provide access to it in new ways. This, in turn, generates new data and information to support other decisions. The integrated nature of the cloud environment, combining data with the means of manipulating those data through the API development platforms, makes that all possible.

With new web-based mapping services, tiled maps have become commonplace for consumer web map services (WMS), such as Yahoo! Maps, Google, MapQuest, and OpenStreetMap. These tiled map services use a REST (representational state transfer) approach for extracting content from a website and are characterized by fast redrawing or refreshing and the ability to provide continuous panning and zooming. REST is the architecture that enables use of any data, not just maps, from multiple locations on the web. As REST is easily accessible through standards such as HTTP and HTML, Esri chose to adopt the REST architecture for making its tiled basemaps available through ArcGIS Online.

In 2010, Esri made its REST implementation an open standard in an announcement at the FOSS (Free and Open Source) conference in Barcelona, Spain. This is expected to have the same standardizing effect on web mapping that publishing the shapefile (.shp) format had on vector geospatial data models during the early 1990s.

Although still under development, the next release of the HTML standard, HTML5, includes native support for multimedia and graphical content and an improved architecture for transferring data from the server to the client. This will obviate the requirement to use other plugins and APIs, and have a significant impact on the delivery of spatial data for time-critical applications.

ArcGIS Online

ArcGIS Online, launched in 2009, is an Esri initiative to create a cloud-based GIS community, a "system for using geographic information everywhere." ArcGIS Online integrates web services and provides a facility for storing and sharing data, applications, and basemaps. This system allows you to discover data and services provided by other users and to share your own data. In so doing, it has the potential to expand existing reserves of spatial data. It is the user community itself that is developing those data resources through a bottom-up, user-driven approach rather than a top-down approach from a government agency or organization. Although Esri created the infrastructure and is hosting the site, the users drive the system. ArcGIS Online provides a kind of social networking for GIS, a "Facebook for GIS users." Users can decide what information to share and who to share it with. This capability defines the access and use constraints for all discoverable GIS data on the system.

Through this system, you can view maps created by others, create your own maps by combining and manipulating online maps, or post data from a desktop GIS. You can choose not to share a map, you can share it with the world, or you can create specific groups and share it with them, in effect, setting up your own data portal. These shared groups could be people in your department, people with common research interests, people in the same geographic area, or other combinations. You can do the same thing with applications. These online resources not only work with ArcGIS software, they are also accessible with a new set of free viewers and open REST APIs. This means that ArcGIS users can share their datasets and services in an environment that is available and accessible to everyone.

As you already know, a GIS is a system. You rarely work with a single aspect of the data such as topology, attributes, geometry, symbology, or some other characteristic of your data. In most cases, you work on a combination of these characteristics that represent the real-world phenomena that you are studying. While it has become easier to fix broken links to files that have become dissociated with the map document, sharing data and symbology with others within or outside your organization has always been a bit of a challenge. To share data with another user, you had to store your datasets on physical media and then post the media. Even after the Internet was generally available, you still had to compress and combine data, symbology, or other files into zipped files which may still have been too large to e-mail. So, you were forced to use an online large-file handling service, an FTP site, or other means. ArcGIS Online provides data structures and an environment for sharing data that helps overcome those challenges.

For example, during 2009, a new data structure known as a layer package was introduced. This was a compressed version of a single or multiple data layers from ArcMap that was made available in such a way that remote users could take data and use them on their computer with the same schema, symbology, and classification scheme created by the data author. One of the most useful aspects of a layer package is the ability to create it from a group of multiple layers, all symbolized and labeled just the way you had configured it in your desktop environment.

In 2010, this was developed a stage further with the release of an additional data structure—the map package. Like a layer package, a map package combined the schema and symbology of one or many spatial data layers, but also added the capability of including the map document (.mxd) in the zipped file. You can easily open a map document and begin viewing the data immediately. ArcGIS Online offers a Javascript-based viewer that allows you to view content with a web browser. The online site also offers a product called ArcGIS Explorer Online, which provides additional tools based in Silverlight that enable you to examine online data. ArcGIS Explorer Online runs in a web browser, with tools to measure, symbolize, query, and hyperlink to data, and includes a presentation toolkit. If the data are also made available as a service, you can view the data inside ArcGIS for Desktop. In 2011, it became possible to drag and drop geographic data, such as zipped shapefiles and CSV files, directly onto the ArcGIS Online site and map their contents. It is also possible to host .zip files online that may include map layers, PDFs, guidelines or procedures, metadata, hyperlinked images, or even videos. This takes sharing spatial data to the next level, as it includes documents that are not part of the data.

It is now very easy to create layer packages and map packages from your data and post them to ArcGIS Online. You can use ArcGIS for Server, cache your data as map tiles, upload them, and make them available as a service. With the release of ArcGIS 10.1, even users without their own ArcGIS for Server software will be able to upload layer and map packages and have them automatically converted into hosted map services using ArcGIS for Server cloud capabilities. These services may then be displayed in the online viewer.

ArcGIS Online can include a variety of data, such as map services, image services, and feature services, as well as map services from OpenStreetMap. Map service publishers who enable WMS on their own computers can register them with ArcGIS Online, so other users can discover and use them directly. In the very near future, Esri will support any WMS service and KML in ArcGIS Online. Esri will also support the display of any WMS service in the ArcGIS Online map viewer and the registration of any WMS service within it. Esri also plans to set up a crowdsourcing service where web users can contribute suggestions for updates or identify issues in the basemaps by sketching in comments. The goal is to forward suggested updates back to participants in the community mapping projects, so they can use the information to make whatever updates they deem appropriate. In this way, VGI and the cloud work together in a mutually-beneficial feedback loop to improve the quality of the data.

The types of data available through ArcGIS Online will also be increasing. At the moment, the focus of the basemap program is on topographic, street, and image maps. However, Esri is considering publishing additional maps, such as parcel, TIGER, hydrology, and other mapped data. For example, in early 2011, Esri and the Natural Resources Conservation Service released a soils web map service containing three soils datasets on ArcGIS Online. What the end-user sees is scale-dependent: At the smallest scale, the NRCS global soil orders data are displayed; at medium scales, the NRCS state-level soils dataset STATSGO (State Soil Geographic) is available; and at the largest scale, the SSURGO (Soil Survey Geographic) data may be viewed. The Esri imagery map service appears in the background, and information on each soil map unit can be accessed, including the name of the soil, the dominant soil order, and whether the soil is suitable as prime farmland. The Soil Photo tool displays a sample photo from the NRCS website of the chosen soil order, along with a soil texture triangle. This information can be compared with other soils in the area.

These developments will become commonplace as more and more key datasets migrate to ArcGIS Online and other cloud-based portals. However, like other data compilations, these collections take time to coordinate, assemble, and post. Many of these maps will be built by the community using the same partnership approach as for the topographic basemap.

Partnerships and collaborations

The cloud GIS environment makes possible some new and interesting collaborations between organizations such as government agencies and industry. In 2010, the European Environment Agency (EEA) and Esri signed a memorandum of understanding to support the design and development of ways to share and access spatial data provided by the EEA's thirty-eight member countries. Those who would benefit from the collaboration are not just in government and industry, but scientists, policymakers, GIS analysts, and the general public. Data sharing would be in line with the principles of the INSPIRE initiative we discussed in chapter 6 and would include standardized templates and layer definitions based on the Esri Community Maps initiative, which we will return to in the next section. The partnership's goal is to provide policymakers with the tools and data they require to analyze environmental issues. The partnership also has the potential to be of benefit in many other areas.

Rather than creating communities of users from scratch, the new collaborations often build on existing ones. The EEA-Esri partnership, for example, builds on the European Environment Information and Observation Network (Eionet), involving 900 experts and over 300 institutions. As Eionet coordinates the delivery of environmental data, it was a natural fit for a spatial data partnership.

DataBasin

Many of the new cloud collaborations incorporate the ability to accept volunteered geographic information. For example, in the environmental GIS community supported by the EEA and Esri that we mentioned earlier, citizens would not only be able to create personalized maps but also contribute their own data. One example of this trend is the new DataBasin resource. Place, GIS, and mapping are seen as integral to solving complex environmental issues. GIS is not just a tool, but has emerged as a common language. GIS is DataBasin's common language-bridging tool and the data are served using ArcGIS Online. As a further reflection of reliance on the cloud and the crowd, its maps are built on a series of Esri basemaps and datasets customized by users, with featured content and direct links to social media outlets. One question the creators of the site asked was, "Could a GIS approach be fashioned that would meet a high scientific standard while, at the same time, appeal to nonscientists?" This resource represented collaboration among the nonprofit Conservation Biology Institute (CBI), the private, philanthropic Wilburforce Foundation, and Esri.

DataBasin is based on six major building blocks: datasets, maps, galleries, people, groups, and tools. Like other portals, the data are seen as only one component of the overall portal, with the people component as probably the most important. Datasets include biological, physical, or socioeconomic themes; they can be uploaded by users, downloaded, or visualized inside DataBasin. Over 1,000 datasets currently reside in the DataBasin warehouse. Users can elect to make datasets they upload completely private, available to specific groups, or available to everyone, reflecting the online functionality of ArcGIS. Resources include galleries, which are collections of datasets, and maps created by DataBasin users. Tools already include an environmental risk avoidance tool, a protected areas planning tool, and a watershed assessment tool. DataBasin also includes physical centers in several locations and with several themes, such as its Climate Center, Aquatic Conservation Center, and its Boreal Information Center, which is currently funded by the Ivey Foundation, Limited Brands, The Nature Conservancy (TNC), and Toronto Dominion Bank. Other planned centers include a Connectivity Center, a Conservation Education Center, and an Aboriginal Peoples Center.

Esri Geoportal Server

In addition, cloud computing also makes for a wider variety and increased interaction between traditional closed-source vendor software and the free and open-source geospatial community. As previously mentioned, Esri made its REST implementation an open standard in 2010. In 2011, Esri moved its Geoportal Extension into a free and open source toolkit called the Esri Geoportal Server, which allows GIS managers to catalog the locations and descriptions of an organization's geospatial resources in a central repository. The repository can be made available to an organizations intranet or to the public Internet. Resources are registered using metadata and the repository helps ensure that everyone in the organization is using the best resources available. The geoportal also gives an organization an enterprise view of its own geospatial resources. Visitors can register their own resources if permission is granted by the organization's GIS managers. Released under the Apache 2.0 license, the Geoportal Server's full source code, documentation, and releases became available on SourceForge, a library of open-source tools. There is no requirement for purchasing ArcGIS Server, and optional technical support from Esri is available. Community contributions to the toolkit are welcomed via the forum, documentation, and the source code. In addition, open government initiatives such as Data.gov and data.gov.uk are driving the development of GeoSPARQL, an open framework for spatial operators and semantic web technologies such as definitions and uniform resource identifiers.

The landscape of modern GIS is therefore marked by a wider variety of public, private, and nonprofit organizations than ever before. Data providers and data users have unprecedented choices on the manner and tools through which they provide those data. Vendors such as Esri have recognized the contributions to the GIS field by the free and open-source community, and are offering some of their tools through that community. More importantly, the field of GIS is increasingly shaped by a wider diversity of perspectives and developments.

Community mapping initiatives

The storage capacity and the ease of programming provided by the cloud environment have spurred initiatives that enable communities to contribute their data to a growing global library of spatial data. Such initiatives have been responsible for mapping local and regional projects as diverse as an Arctic Bay Spoken Map (with over 300 traditional Nunavut place names), a City of Memory map (funded by the National Endowment for the Arts and The Rockefeller Foundation to document New York City experiences), and hundreds of environmental maps of communities worldwide. Community mapping initiatives have their roots in *Participatory GIS*, which began in the 1980s. Growth was initially modest in this area until the advent of easy-to-use map creation tools and improved network data transfer rates that allowed maps to be showcased online. One large community map project is the Aboriginal Mapping Network, which was established in 1998 as a joint initiative of the Gitxsan and Ahousaht First Nations and Ecotrust Canada. Indigenous technicians and decision makers now practice and share traditional knowledge and use GIS mapping techniques and the Internet to investigate issues such as land claims, treaty negotiations, and resource development around the world.

Esri Community Maps Program

It has been difficult for interested government agencies to collaborate at different levels with a view toward integrating and harmonizing their datasets into a continuous map.

The barriers to collaboration have been organizational, technical, and financial. This is true both at the national level and within the same region or county. One metropolitan area may have dozens of local governments and authorities, each facing budgetary limitations and staff shortages, making it difficult to even collaborate among their own departments. This lack of collaboration plus the advent of increased demand for authoritative sources on the cloud prompted Esri to establish the Community Maps Program during 2010.

The Esri Community Maps Program provides a way for organizations to contribute their data to the cloud. Data sought for the program includes imagery—at spatial resolutions finer than one meter and collected in the past three years—to improve what is already in the current World Imagery database on ArcGIS Online. The program also seeks street and basemap data at a scale of 1:50,000 or higher in countries where these data are not already available; local basemap data at 1:5,000 scale or larger, such as buildings, trees, sidewalks, and other high-resolution features; and university campus basemaps. Contributed community data are integrated with data from other providers and then published through ArcGIS Online as a map service. Users will be able to use the mapped data through ArcGIS, ArcGIS Explorer, ArcGIS for Server web mapping applications, or even a web browser via ArcGIS Online. The only stipulation is that the data must be provided in a file format and coordinate system supported by Esri software. To ensure uniform, high-quality cartography, map templates are provided to communities and university campuses. Once the map has been authored, it can be used to generate a map cache either by the user using ArcGIS for Server or by Esri. Esri then publishes the map cache and integrates it into existing data online.

This program required a new enabling technology, a common multi-scale map template, and a new process for integrating basemap data, all of which comes together online. Community basemaps are collaborative and multi-scale. They also come from an authoritative source, such as a local or regional government agency, that is already creating data as part of a transportation department, a planning department, a public works department, or other existing operation. The basemaps from each community are served on ArcGIS Online in a seamless format along with those from other communities. Content is contributed according to the template, uploaded as tiles, and served in the form of map cached tiles. The basemaps are open and accessible through the public REST API. In early 2011, it was announced that all basemaps published and hosted by Esri are freely available to all users regardless of commercial, noncommercial, internal, or external use.

There are three ways to use and license community basemaps. They can be used as free map services for integration within local GIS systems, or they can be used as a foundation for building web applications (mostly public sector-oriented but also commercial applications). They can also form part of a *data appliance*, a combination of hardware and a customized bundle of data. Online ArcGIS basemaps, along with other layers, including commercial datasets, are packaged onto pluggable hard drives for easy deployment within an organization's existing networking and security infrastructure. The public content available with the community basemaps program is provided at no additional cost as part of these appliance configurations.

Many organizations will find it advantageous to meet the demands of their users and their customers by placing some data in the cloud. The Esri Community Maps Program has proven to be a popular innovation, with over 1,500 participants signed up for the first series of training sessions during the summer of 2010. Cloud-sharing initiatives like this have great potential to expand the breadth and depth of public domain spatial data available to GIS users. Agencies download a map template, add their data to the template, insert their map tiles, and gradually build a multi-scale map of the world.

Beyond data sharing, the Community Maps Program is also enabling different governments and other organizations to work together across administrative and physical boundaries by building data resources around common templates and requirements. Some international agencies that charge for their data have participated in the Community Maps Program to publicize the benefits of their data. They are, in effect, using the program as a shop window to promote their data products. As users cannot download data from the basemap tiles on ArcGIS Online, the placing of data online is not seen as a practice that would cut into data sales.

DataDoors and other cloud services

Many other cloud services currently available have a spatial component. For example, as we mentioned in chapter 3, i-cubed runs DataDoors data management products. These products offer a close integration with ArcGIS for Server and support the discovery and delivery of raster and vector data. They enable the user to search, via secure Internet connections, local and remote, public and private archives of geospatial data such as satellite imagery, aerial photography, topographic maps, and digital elevation models (DEMs) from within ArcGIS for Desktop software or via a web browser. DataDoors then delivers information about available coverage and allows users to view and retrieve relevant data customized to the user's specifications.

Tools like this reflect the growth of the cloud environment for applications as well as data. DataDoors, for example, allows you to include your own algorithms or GIS and remote sensing processes, such as orthorectification, the creation of vegetation indices, change detection analyses, and land use and land cover classification techniques. The DataDoors Process Manager extension even helps manage the workflow itself.

C/JMTK Geospatial Appliance

Other geospatial cloud-based services are tailored to the defense and intelligence communities. For example, the Commercial Joint Mapping Toolkit (C/JMTK) Geospatial Appliance was built by Northrop Grumman Corporation, a defense contractor, in collaboration with Esri. The appliance (a package of hardware, software, and services) delivers basemaps for defense and intelligence operations, combining classified datasets from the US National Geospatial-Intelligence Agency with commercial datasets.

Data Hub

Similar in some ways to ArcGIS Online, Data Hub, part of the Comprehensive Knowledge Archive Network (CKAN), is composed of groups of spatial data. It is a part of the Open Knowledge Foundation Limited, a not-for-profit organization founded in 2004 in the United Kingdom dedicated to promoting open knowledge. The site lists free and open access to the material, freedom to redistribute the material, freedom to reuse the material, no restrictions based on who someone is (such as their nationality) or their field of endeavor (such as commercial or noncommercial) as the guiding principles behind open knowledge. Unlike many other online resources, Open Knowledge Foundation data are available for commercial use.

GeoCommons

FortiusOne, Inc. developed the GeoCommons community website in the wake of the Hurricane Katrina disaster that highlighted the inadequate response times of traditional map-making cycles and the frustrating delays in getting the most current and accurate data to relief agencies. This community website allows anyone (subject to registration with the site) to contribute data, build maps, and share data, as well as download data and maps contributed by others. The GeoCommons public data portal has been developed on the GeoIQ cloud; a secure cloud platform that offers data management, visualization, and analysis for both government and commercial, technical and nontechnical users. Whereas GeoCommons is all about open public data, the focus of GeoIQ is private data.

Pros and cons of using and hosting spatial data in the cloud

As with any other technical innovation or conceptual issue in GIS, you need to think critically about cloud computing so that you understand its benefits and limitations. One major concern for anyone hosting and using data in the cloud environment is security. In an era where security breaches are commonplace—from Google Docs containing sensitive information being accessed for unlawful means to stolen laptops and hacked databases at financial institutions—data managers have good reason to be wary about data security, including when and where data are transferred, stored, and used internally and externally. Given the vast amounts of information maintained in GIS databases worldwide, how can data be recovered in the event of disaster and how can data be legally protected? Users of these cloud services must be assured that all appropriate safeguards are in place for disaster recovery, business continuity, and the availability of audit trails and transaction logs. In addition, a robust and reliable payment framework must be available, as well as the protection of intellectual property rights and guarantees of liability.

The Open Geospatial Consortium, Inc. (OGC) is working to develop standards for geospatial technology and services in cloud computing. The OGC acknowledges that cloud computing may be difficult in a secured or classified environment where security is paramount and where computers must be stand-alone and not networked. Security issues may also be present at the firewall level and, therefore, some geospatial data or services may not be shared in the cloud due to security restrictions. The OGC is advancing best practices for implementing standards on security in the cloud through its Interoperability Experiments. It has also developed GeoXACML (Geospatial eXtensible Access Control Markup Language) to facilitate this development. The OGC also supports a spatial law and policy committee, as well as a workflow domain working group that addresses geospatial workflow, including security and licensing issues such as data encryption, authentication, and provenance tracking. Perhaps most importantly, the OGC is working on procedures that will one day enable reviewers and users of spatial data to seamlessly inform data publishers of changes that need to be made in their database.

Closely tied to the issue of cloud security is privacy. As you read earlier in this book, privacy is of particular concern in GIS for several reasons. These include the ease with which personal and residential information may be related, the increasing availability of very accurate and current imagery recording what is happening on the ground, and the large amounts of often sensitive, detailed information—ranging from the location of military test sites to the demographics of individual homeowners—that may be contained in a GIS data source. As a result, some organizations are choosing to establish a private cloud. This array of servers and applications exists behind an organization's firewall and are maintained within the operating parameters of that organization. What could emerge is a mixture of secure private and community clouds, each serving the specific requirements of individual organizations and data users in some way.

Another concern is ownership of data once the data are placed on the cloud. Who owns the data when an organization or individual places their data on someone else's server? For online content, the end-user agreement Esri has adopted is a version of the creative commons license agreement that stipulates free use of the data for noncommercial purposes. Data cannot be resold and the contributing organization or agency is provided with full attribution that they created the data and retain ownership. When reviewing terms from data providers, Esri looks to ensure that the providers will support this end-user agreement. The philosophy behind the Community Maps Program is to provide mapping for everyone. The basis of agreements between Esri and state, local, and national agencies is that the agencies providing basemaps and imagery data will in return receive free web services for their content. This is in contrast, at the time of writing, to the Google terms of agreement, where the data hosted becomes the property of Google, and not that of the organization where the data originated.

Esri's goal is to minimize the time and resources required to execute these agreements, particularly on the side of the organization or agency. Typically, Esri's approach first leverages the existing usage terms published by the data contributor, often as part of its data clearinghouse. Secondly, a simple participation agreement is executed, granting Esri the right to host and serve the basemap to anyone under a creative commons license agreement. To date, most government contributors have accepted these terms unconditionally. In some cases, government organizations have developed specific terms for sharing their data and have asked that Esri adopt these terms. The terms refer to the specific laws, mandates, and other requirements that are adopted as part of their stated policy. Esri has, in turn, now incorporated several of these terms into its standard agreement.

The basemap services are provided free to all data users for noncommercial use. Fees will be generated through licensing Esri's API, thereby recovering the cost of hosting and serving the information for commercial applications. Esri does not obtain ownership or IP rights to content shared by community basemap users. In all cases, the contributing organizations retain full rights to their shared data. Furthermore, Esri retains no ownership rights over other data that are shared on ArcGIS Online, such as layer packages and map packages. All ownership is retained by the originating or contributing organization.

Although a great deal is made of the robust and resilient nature of cloud computing, no service provider can guarantee 100 percent availability. Power failures, technical faults, terrorist attacks, volcanic eruptions, or even increases in solar flare activity are among the many potential threats to cloud services. Other problems have arisen when data seem to be lost after being placed online, which seems to be happening more frequently. In one widely-publicized case, Sunrise, Florida, a city of 90,000 people, disappeared for a month during the summer of 2010 (Sutter 2010). As this happened on Google Maps, it posed no serious problem to the management of the city, but the disappearance inconvenienced tourists and locals alike who couldn't locate services. As a result, it was reported that business revenues and the city's sales tax earnings declined. The mayor said that the same thing has happened in Sunrise (twice before) as well as in other cities. If a large amount of critical infrastructure and public safety spatial data suddenly disappeared from a cloud environment, this would have far-reaching implications.

For these reasons, the hybrid cloud implementation, with a combination of cloud-based and local resources to support communities and organizations, appeals to many. This allows users to take advantage of all the cloud has to offer but at the same time provides a safety net and backup resource in the event that users cannot connect to the cloud for whatever reason. Not having access to a cloud service should not mean everything grinds to a halt.

Esri designed its online ArcGIS site with appropriate redundancy to keep systems from going dark in the event of emergencies. The cloud services that make up ArcGIS Online, and all of the GIS services and data that they reference, are already spread across the Internet. For the content Esri hosts, such as the community basemaps, this has been replicated across distributed data centers and cloud-computing farms across the world, as well as in Data Appliance for ArcGIS installations at multiple federal agencies.

Anyone wishing to submit their spatial data to the cloud should think carefully before they do so. Do you or your organization have the right to post the data you are using online? This often depends on where the original data came from, who created the data, what the reuse restrictions are, and what processes were subsequently run on the original data which may have resulted in data alterations. For example, if an organization wants to participate in Esri's Community Maps program, the data it contributes must be royalty-free and able to be redistributed to anyone. That is, the data must be in the public domain. The data also must contain basic metadata so that the source is documented and the content may be searched for. Many organizations work with proprietary data that they themselves license or use data that they have purchased from others. As a result, they are unable to share those datasets with others. In some organizations, simply identifying the proprietary data would be an unjustifiable use of limited staff resources. This may be especially difficult to do in organizations where the original data have been edited by different departments over a long period of time, using a variety of procedures, some of which may not be well documented. In other organizations, such as tribal governments that maintain data holdings on sacred sites or city governments with information on property owners, the data are deemed too sensitive to release. Finally, in some organizations where data have only been shared internally for years, metadata availability is patchy at best.

Alongside the recent developments in cloud computing is an increased desire by those in the GIS community to share their data with those outside that community. Network data transfer rates have now improved considerably and allow live maps to be streamed, accessed, and customized to specific needs. Servers have become cost-effective, proliferated in terms of number and storage capacity, and now support organizations storing huge amounts of data. APIs have developed to the point where the programming to serve maps and data online has become relatively straightforward. As a result, the maps generated by GIS software and services can be made available to new audiences—from policy makers to ordinary citizens—for GIS services. An illustration of this is found at Data.gov, where a GEO Viewer tool allows any users to preview geospatial data available through the Data.gov catalogs. GIS users can use this viewer to assess whether datasets meet their requirements before downloading the data. Non-GIS users can overlay a series of maps to illustrate and understand certain phenomenon. One of the data layers available on the site, the location and attributes of world copper smelters, is shown in the following:

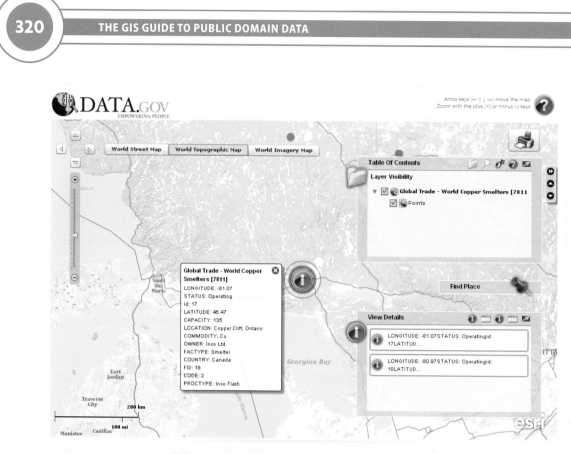

Basemaps available from Data.gov from ArcGIS Online services. Courtesy of Data.gov.

As discerning GIS users, we need to distinguish between resources that provide access to online maps and resources that allow us to interrogate the data and derive our own information products from those data. In the spectrum of available tools, each has its place, but if your goal is to analyze the data, a data viewing utility like the Data. gov example is unlikely to provide the necessary analytical capability. Very few of these online maps have a download raw data option, because the goal is not to support the GIS community as such, but rather to provide general information for the non-GIS community and public at large. To interrogate the data further, you would need to investigate the GIS data section of Data.gov to analyze copper smelters by themselves and in the context of other data, such as rail, road, and river transportation, mineral belts, bedrock, and surficial geology.

As with any resource new to the GIS scene, ArcGIS Online offers great potential but does have limitations and is changing rapidly. The online service was established to allow GIS data users to share information. Although users must register with the service to post data online through an Esri Global Account, the possibility remains that some of those data may be inaccurate, incomplete, or of poor quality. Some data may not be documented well or at all. Old and out-of-date versions of a dataset may be uploaded. Even worse, individuals may upload copyrighted data but fail to highlight the restrictions on use for other users. At the moment, the documentation section adopts an honor system, and it is up to the user to populate it. Soon after the restructuring of the online ArcGIS (2010) enabled users to post their data to the site, the site was populated by thousands of data items. Some of those items belong to specific groups, but little of the data is in clearly defined and labeled folders. Instead, the data available online feel a bit like an unstructured library. Finding data in this library relies on searching, with search results only as good as the metadata and tags included in the data. Once again, it all comes down to the metadata!

LOCAL DATA FILES

Another thing that you should be aware of when using cloud environments is whether the data you are using are actually being downloaded and stored on your local hard drive. For example, when you download data online from ArcGIS, the data are unpacked using a pointer file called pkzip.info. This pointer file retrieves the necessary data and saves the data on your computer in a folder it creates under your user name under Documents and Settings. (Note: The name and location of this default folder may vary depending on your operating system.) As this book has emphasized, a key skill in success with GIS is being a good data manager. This default folder, buried in a series of other folders on your computer hard disk, is rarely a good final home or working folder for your data. If you work online with ArcGIS for long, you might want to keep an eye on this folder to make sure it does not become so full that you run out of hard disk space. You should move data saved to this folder to a more suitable location and make certain that the map document (.mxd) is set to use relative paths.

Cloud-based services make it easier than ever for GIS users to obtain data. However, it's also easier to bypass the critical step of becoming familiar with the data you are using and the procedures and decisions used to create the data. Although in some cases it is a difficult, often painful process, when you are forced to manipulate a dataset to bring it into the correct format for your GIS, the hours or days you spend forces you to get to know that dataset, both the good and the bad points. Cloud-based services reduce the time that you must spend with the data before the analysis stage, but it's also more likely you will proceed to use the data before you fully understand the data. For example, the USGS StreamStats web GIS significantly reduces the time needed to prepare hydrologic data. StreamStats hydrologic information is generated from datasets and procedures that are unseen and available behind the scenes. Some of the data used includes DEMs, which have specific resolution, precision, and limitations imposed on them when they were created. They are also based on topographic maps or remotely-sensed imagery, both of which had accuracy tolerances and limitations. StreamStats also uses procedures, such as regression equations, which have specific parameters such as peak flow, drainage area, basin elevation, and mean annual precipitation. While a goal of many GIS projects is to reduce time spent to prepare data and increase time available to analyze data, it is still important to understand the data so that your results will be appropriate, relevant, and accurate. Even though it is possible nowadays to just plunge right in and start using data from a myriad of sources, resist that temptation; know what you are working with. Your decisions are only as good as your data.

Reflect upon and answer the following questions about cloud computing's connection to GIS:

Does cloud computing really remove the complexities of GIS? How does it do this? How might it add different types of complexities?

Does cloud computing make GIS more open? How so?

Does cloud computing obviate the requirement to become an expert GIS user? If so, does this introduce any other problems?

In terms of ArcGIS Online, much that was cumbersome about sharing and using GIS data has been removed. However, GIS is still a system, part of a larger discipline of GI Science, bound up in its own terminology and best practices. You still need to think critically about how you can best position yourself to take advantage of new methods and tools, including cloud-based technologies.

Web versus desktop GIS

Despite the great technological changes in GIS over the past thirty years, data workflows have remained remarkably constant. While it is true that more spatial data have become available and accessible, and that hardware, software, and computer networks have certainly became more powerful and mobile throughout those three decades, the hardware-software model in GIS use changed little. GIS, as a desktop software application, was installed and run locally, accessing data from external devices such as magnetic tapes, zip drives, CD-ROMs, DVDs, and so on. As the technology evolved, data became increasingly available to download from the Internet, although invariably still stored on a local hard disk or network storage device.

This model survives in most GIS implementations today. However, with the advent of web-based GIS around the year 2000, methods available to learn about and use GIS have evolved. These changes present opportunities as well as challenges. As never before, data users have a wealth of geospatial tools at their fingertips. Web-based GIS tools can help you visualize patterns, trends, and relationships among spatial data. Dynamic content, such as real-time wildfires or earthquakes, can be analyzed online in a way that is not possible with traditional downloaded data layers. Web GIS sites provide some basic analytical tools for exploring and solving problems using public domain spatial data. No software is required beyond a web browser, and most web-based GIS services require little time to learn. An advantage of web GIS over desktop GIS, particularly in countries where all spatial data are licensed, is the data in web GIS are accessible and viewable, although in many instances, using the data will incur costs or licensing restrictions.

Challenges also exist in using web-based GIS to explore public domain spatial data. Web GIS sites are usually focused on a particular theme, making it difficult for you to integrate data representing other themes. For example, a web GIS site focusing on hydrography will probably not allow population data to be added. You are usually prevented from using your own data. Even if you are allowed to use your own data, you will be restricted to certain file types and may only map a small amount of data. Information about data quality on web GIS sites is usually more difficult to ascertain due to incomplete or missing metadata. The quality of the data, like other resources on the Internet, ranges from abysmal to excellent. The use of web GIS, like other online tools, introduces additional data storage and security concerns. All of these shortcomings are gradually being addressed in the current rapid development and deployment of increasingly powerful web GIS tools. At present, the chief disadvantage of web GIS for solving problems is the limited functionality available compared to the advanced toolsets available to the desktop GIS user.

However, as a GIS problem solver, you are not confined to the choice of either desktop GIS or web GIS. With recent developments, you now have the hybrid option of using desktop GIS but streaming data from web-based GIS mapping services. This provides access to the full suite of desktop GIS tools but obviates the requirement to store large amounts of data locally. Updates are done on the server side and you simply access these datasets through online services and tools, from servers using ArcGIS for Server, including the online ArcGIS, or a web mapping service. In the following activity, you will have an opportunity to compare web GIS, desktop GIS, and a combination of the two using the National Atlas of the United States.

COMPARE WEB GIS WITH DESKTOP GIS FROM THE US NATIONAL ATLAS

Access the National Atlas of the United States on http://nationalatlas.gov.

How many federal agencies are involved with hosting data on the National Atlas?
Which participating agency is responsible for maintaining the National Atlas portal?

Click the Map Maker button and make online maps of various themes of your own choosing. You will see a great many choices here, literally from A to Z (agriculture to zebra mussels). When you have made a map, select Refresh Map and then select Map Legend at the top of the layer list. For example, you could make a map of cotton production, crime, or the distribution of the invasive species, purple loosestrife.

Did you see any patterns on your maps that were surprising? If so, what were they?

Now that you have used the National Atlas as a web GIS, you will next download a layer from the atlas and use it in your desktop GIS. Begin by selecting any of the hyperlinked names in the layer list. You will be directed to the data description section. Once there, you will see a download raw data choice (in the lower right). Upon selecting it, you will be placed in the data download listing. Make sure this is a vector, not a raster dataset. If you have a choice of downloading an Esri shapefile or a SDTS file, select the shapefile.

Recalling chapter 1's discussion of SDTS, why should you avoid using SDTS if you have a choice of an Esri-ready format?

After downloading your layer, unzip the file and add it into a new ArcMap map document.

In a few sentences, reflect upon your experience of the ease or difficulty of downloading data from the National Atlas.

What are the advantages and disadvantages of mapping public domain data online using the web GIS service in National Atlas versus downloading it and using it inside ArcMap?

The National Atlas also provides a way to stream data from its web mapping service to the desktop GIS environment. To do this with the National Atlas, in ArcMap, select Add Data. Browse all the way to the top of the directory tree and Add Web GIS Service. Add the URL http://nationalatlas.gov, and add one or more of the layers that you see listed, whether it is rivers, agricultural census data, crime, or another theme.

What is the advantage of using this third method, desktop GIS together with web-based spatial data, over the first method of using web GIS with an Internet browser?
What is the advantage of using desktop GIS together with web-based spatial data over the second method of using desktop GIS with local data?

The National Atlas certainly is a wealth of spatial data, representing over 200 layers. It is an excellent example of a national mapping portal that serves the needs of the general public as well as GIS users. National Atlas data can be used to solve a variety of problems using public domain spatial data.

Summary and looking ahead

In this chapter, we have investigated the advent of cloud computing—what it means and how it is being implemented. We looked at some early adopting agencies and programs, and considered some of the pros and cons. Now that you understand what cloud computing is and how it can affect the work you do with GIS, you are ready to apply your new knowledge to solve some real-world problems.

In the next and final chapter, we will look at some of the emerging sources of public domain data and some of the new tools and methods that will become increasingly available to access and analyze those data sources. We will revisit some of the issues discussed in earlier chapters, such as data quality and fee vs. free, and discuss what the future may hold for GIS and public domain data.

References

Hemsoth, Nicole. 2010. "GIS Applications Take to the Clouds." *HPC in the Cloud* e-magazine (July 20). http://www.hpcinthecloud.com/hpccloud/2010-07-20/gis_applications_take_to_the_clouds.html.

Higginbotham, Stacey. 2010. "Sensor Networks Top Social Networks for Big Data." *Gigaom* (September 13). http://cloud.gigaom.com/2010/09/13/sensor-networks-top-social-networks-for-big-data/.

Mell, Peter, and Timothy Grance. 2011. "The NIST Definition of Cloud Computing." *National Institute of Standards and Technology*. http://www.nist.gov/customcf/get_pdf.cfm?pub_id=909616. NIST Special Publication 800-145.

O'Doherty, Philip. 2010. "How Will GIS Companies Weather the Cloud Computing Storm?" *GIS Lounge* (April 29). http://news.gislounge.com/2010/04/%E2%80%9Chow-will-gis-companies-weather-the-cloud-computing-storm%E2%80%9D/.

O' Doherty, Philip. 2010. "Every Cloud Has a Silver Lining; Only 'GIS as a Service' has Potential to Meet Growing Demand for Geospatial Technology." *Directions Magazine* (September 8). http://www.directionsmag.com/pressreleases/every-cloud-has-a-silver-lining-only-gis-as-a-service-has-potential-to/131994.

"RMSC and Esri Collaborate on GIS Cloud." 2010. *Inside HPC* (July 12). http://insidehpc.com/2010/07/12/rmsc-and-Esri-collaborate-on-gis-cloud/.

Sutter, John D. 2010. "Google Maps 'Loses' Major Florida City." *CNN Tech* (September 22). http://www.cnn.com/2010/TECH/web/09/22/google.lost.sunrise.florida/index.html?iref=NS1.

THE FUTURE OF PUBLIC DOMAIN DATA IN GIS

Introduction

Now that you've pondered issues and solved problems using public domain data in a GIS, one key theme should be evident: GIS has experienced rapid development from its inception to the present. The types, formats, and means of delivery of the data that fuel GIS are undergoing more change than ever before. We have investigated several new developments in this book, including spatial data infrastructures and standards, cloud computing, and volunteered geographic information. Many of the forces acting as agents of change on GIS also exist within the GIScience and data communities. In this chapter, we will examine recent initiatives and emerging issues in GIS, with our focus on those that affect anyone wishing to use public domain spatial data. In doing so, we encourage you to envision the kind of future development you would like to see in the field of public domain data and GIS. Work with the GIS community to make your vision a reality.

New public domain spatial data

New sources of public domain data are becoming available every day. Given the ease of setting up a data portal and the vast reserves of spatial data that have accumulated within many organizations, this is hardly a surprise. Some of these sources are long-established mapping and land management organizations, some are organizations new to the geospatial scene, and some are new partnerships between old and new organizations. Organizations, such as nonprofit agencies involved in land use or environmental issues and not previously in the business of providing access to data, are increasingly doing so. One example is The Nature Conservancy (TNC), which took advantage of replacing its map server in 2010 by offering new map services (in both geographic and Web Mercator projections) based upon its core conservation data. The services include assessments and portfolios for ecoregions, conservation projects, and landholdings. TNC also hosts a number of services from around the world, such as marine protected areas in Indonesia. These map services are available for ArcGIS, Google Earth, and ArcGIS Explorer users. Web map authors can consume these services either as WMS or ArcGIS for Server formats. TNC's data access policy is to make some data available to the public and to make unfiltered versions of the data, including sensitive records, available upon request. Some of the requests are sent through a review process before being approved.

New data, in both raster and vector formats, are also available from state and federal governments and private companies. For example, a new seamless, national land cover vegetation dataset was released by the USGS in 2010. The dataset, containing 551 ecological systems and thirty-two land use classes, was the result of the twenty-year Gap Analysis Program (GAP). This dataset forms the basis of studies determining the status of biodiversity, assessing climate change impacts, and predicting the availability of wildlife habitat. This particular resource reflects another trend in spatial data provision—collaboration. In this case, the USGS collaborated with the University of Idaho to make the data available via the university's website. The map viewer displays data at multiple scales, eight classes at level 1 (grassland, shrubland, forest, aquatic, sparse and barren, recently disturbed, riparian, and human land use); forty-three classes at level 2, along with elevation and climate information; and 583 classes at level 3. Users can calculate statistics on the types of vegetation occurring within a mapping zone, a state, or a county. Similar viewers also illustrate the increasing availability of improved analytical capabilities through a standard web browser.

Two other trends identified in this book, standards and partnerships, have also played a part in the development of the GAP dataset. Standards helped make the new dataset possible, as the GAP national land cover data are based on the NatureServe International Terrestrial Ecological Systems Classification. Partnerships with other organizations helped fill in missing details; in parts of the continental United States where ecological system-level GAP data had not yet been developed, data from LAND-FIRE (the Landscape Fire and Resource Management Planning Tools project) were used. LANDFIRE is the result of a five-year project between public agencies (the US Department of the Interior and the US Department of Agriculture [USDA]) and TNC. This project, and many similar projects, also reflects the trend to provide data in a seamless, integrated manner. GAP generated state-by-state land cover data beginning in the 1980s, followed by regions in the 1990s. This initiative is also part of a wider context, the National Biological Information Infrastructure.

Another example of new data is an increasing variety of satellite imagery. The Geostationary Operational Environmental Satellite (GOES) program is a partnership between National Oceanic and Atmospheric Administration (NOAA) and NASA. NOAA manages the program, establishes requirements, provides all funding, and distributes the data, while NASA procures and manages the design, development, and launch of the satellites for NOAA on a cost reimbursable basis. NASA and NOAA's latest satellite, GOES-15, came into operation during 2010. The GOES fleet of geostationary environmental weather satellites provides imagery for both the earth and the sun and tracks life-threatening weather and solar activity that could have a detrimental impact on the satellite-based electronics and communications industry. GOES-13 operates above the eastern United States and GOES-11 operates in the western United States. Both provide weather observations covering more than 50 percent of the earth's surface. GOES-14 and GOES-15 operate at 105 degrees west longitude, near the center of North America. Such partnerships between agencies, and between public- and private-sector organizations, will become increasingly common due to the high costs associated with these programs, ever-tightening budgets, and the increasing demand for the data collected. In an innovative move to exploit the broadcasting power of the Internet, the first GOES images were actually released via the University of Wisconsin—Madison Cooperative Institute for Meteorological Satellite Studies' *CIMSS Satellite Blog.*

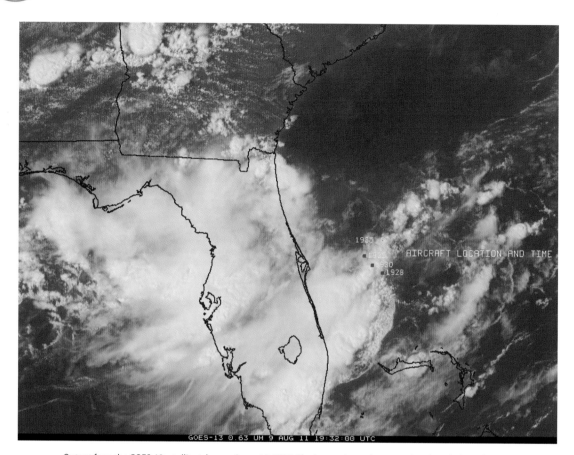

Output from the GOES-13 satellite taken on August 9, 2011. The image shows the route aircraft took through an area of turbulence east of Florida. Courtesy of Cooperative Institute for Meteorological Satellite Studies (CIMSS), University of Wisconsin–Madison.

Finally, detailed data are becoming available for regions of the world that were previously only covered at the coarsest of scales. One example is in Russia. The Russian Federal Service for State Registration, Cadastre, and Cartography (Rosreestr) portal (based on ArcGIS for Server) planned to host over eighty million land parcels online, topographic maps down to 1:100,000 scale, community basemaps at 1:10,000 scale for many cities, and possibly submeter imagery by the end of 2011. While it is true that downloadable data for many areas of the world remain extremely limited, viewing spatial data online is becoming more commonplace.

Partnerships in public domain data

As we have explored in this book, many countries around the world are moving towards more open access to both spatial and nonspatial government data. While this has certainly not been ubiquitous, and in many areas progress has been slow or nonexistent, in general the momentum behind the open data access movement shows no signs of abating. Many governments are also no longer content to simply publish data for the GIS community to download. They are extending the audience to include non-GIS professionals and the general public by making their data available as map services. Despite the relative ease in using GIS nowadays compared to thirty years ago, it is still difficult for non-GIS professionals to turn government data into usable information. A map service, on the other hand, makes it possible for more people to do what they have done with paper maps for centuries—determine patterns and relationships themselves, without relying on the interpretation of others.

The technology behind web-based GIS has encouraged open government and the engagement of citizens with their governments; citizens now have the means to hold authorities accountable and participate in the planning process at both the local and national level. These efforts also encourage government agencies to collaborate with each other on issues that transcend political boundaries. Esri President Jack Dangermond, speaking to *Information Week* magazine, said,

> While I am very excited about what Data.gov has accomplished, I think we need to now evolve our thinking from "Data.gov" to "understanding.gov." This would require more thinking and some additional work on the government side (or perhaps on the private sector side) but the results will be enormous . . . (*Information Week* 2010).

When it became clear to those in GIS that the demand for government-sourced GIS data would outstrip the means of government to provide the data, private companies set up data portals that hosted a combination of their own and government data. In many cases, the portals provided the same information that was available on government sites, but they made it easier and faster for users to access and download, plus they provided additional choices, such as obtaining the data in different map projections. We have already looked at several resources that were originally based on government data, such as DataMart and GIS Data Depot.

Increasingly, these tools are lightweight web-based viewing clients rather than large portals. They also tend to offer more than just data, such as services that connect citizens with government programs, activities, and elected officials. Mapping tools also tend to be created by nongovernment agencies. One example is GovMaps.org, created by Esri in collaboration with government agencies, which invites users to discover hundreds of maps and datasets. Another example is OpenGovernment, which features learn, track, and share tools for citizens who wish to stay in touch with their state. OpenGovernment was created by the Participatory Politics Foundation and the Sunlight Foundation. Increasingly, mapping tools are being seen as an integral part of open government initiatives. Why? Perhaps it is partly because, as discussed in chapter 1, maps have long been seen as sources of information that not only inform their users, but also empower them.

During the first decade of the 2000s, private industry played a greater role in providing spatial data, from satellite imagery to vector data, along with value-added services for government data. This led to a number of private-public partnerships being established, such as the recent work undertaken by Esri and the US federal government to improve the Data.gov resources. Esri will improve Data.gov's mapping capabilities, register new map services on the site, and merge the Data.gov and Geodata.gov initiatives. This will also help extend geospatial analysis beyond the realm of geospatial professionals. Decision makers will be able to create their own maps with ArcGIS Online, combining sets of statistics in the form of maps or charts to illustrate spatial relationships.

For example, it would be possible to overlay freeways, neighborhood housing values, and vacancy rates to determine the impact of highways on urban neighborhood economics. Users could access a map layer representing revenue from fishing and related activities, unemployment rates for the summer of 2010, along with an outline of the Gulf of Mexico oil spill to determine if any relationships exist, where they exist, and to what extent. This is in keeping with federal initiatives to encourage managers to use new technologies to foster transparency, collaboration with industry and governments, and public participation.

Starting in 2011, ArcGIS Online offered historical and current Landsat imagery for the entire planet. This supports monitoring land use and land cover change as a result of diverse influences including agriculture, glacial retreat, mining, and urbanization across a forty-year time span. This would have been inconceivable a short time ago, in terms of the amount of the data being offered, the fact that the data are now free, and the public-private partnership that was required to make it happen. Not only are the raw data offered for GIS users but, reflecting the trend of offering spatial data to the general public, a change viewer is also offered. A historical Landsat scene is shown alongside a current scene for any area of the earth's surface, as well as a land cover image that indicates the change in that scene.

Resources such as GovMaps.org and the European Open Government Data Initiative (govdata.eu) could be considered the first ventures into *GIS for the masses*. Could it be that federal agencies are realizing that the one-stop portals are still too difficult for those outside the GIS industry to use? GIS professionals, and increasingly the GIS community, may be moving away from portals, not to them. However, despite the increasing utility and functionality of these GIS-for-the-masses resources, many are still largely set in stone. That is, they are best suited for the general public to visualize data on maps. However, GIS users cannot download the data, nor can they change the maps or add and delete content. The one major exception, so far, is ArcGIS Online. The user can interact online with ArcGIS using the more powerful desktop GIS tools. How will the different needs of the GIS users and the general public influence this online landscape of data and maps?

Esri also introduced an online product called Community Analyst, which integrates many different local, state, and federal spatial datasets into a web mapping application, making those datasets available as a simple application for government agencies, NGOs, and others to use for planning and community analysis. The Community Analyst site integrates demographic data, population forecasts, housing information, and economic and social policy data, and presents those data in the form of maps. It is similar in concept to Business Analyst Online, which Esri manages for the private sector to support access to business, consumer, and demographic information. Community Analyst targets public-sector data agencies and NGOs with public domain data. Community Analyst exemplifies trends evident throughout this book; namely, a combination of public and private datasets, offered by private industry, packaged for users who have traditionally not used GIS, and served online.

Well into the second decade of the 2000s, two major trends in collaboration are emerging, one at the global level and the other at the regional and national level. First, it was increasingly apparent that while many countries funded national mapping efforts, analysis at the global level remained difficult. Many social, economic, and environmental issues do not stop conveniently at national boundaries, yet little information was available on both sides of those borders to support a more holistic approach to studying cross-border issues. You may have experienced this to a degree when using paper topographic maps. The area you were interested in was inevitably along the edge of two maps or even at the corner of four maps. This very issue was the driving force behind such projects as Global Map, mentioned earlier in this book.

Second, at the regional and national level, many governments faced a shortage of revenue to cover costs of services that were increasingly in demand. News of government bailouts and federal systems saddled with enormous debt became all too familiar.

Given the high costs involved, partnerships to generate spatial data will only increase under such fiscal constraints. For example, late in 2010 the State of Colorado sought partners to cooperate with the USDA's Farm Services Agency to generate new National Agricultural Imagery Program aerial photographs. To fly and produce the data would cost about $170,000 for 4-band color imagery at 1-meter resolution.

Along with shrinking budgets, many governments have been compelled to reduce field staff. This makes crowdsourcing a potentially attractive alternative resource for at least some of the field work no longer undertaken. It is likely that the growth of diverse volunteered geographic information, provided through formal arrangements with government agencies or generated in a viral fashion through social media, will alter the entire landscape of public domain spatial data. It's not difficult to foresee a time in the present decade when traditional authoritative sources of spatial data shrink to a small percentage of available spatial data. It's also not difficult to foresee that shrinking budgets will make rigorous cross-checking of the incoming volume of spatial data next to impossible. Therefore, developing your skills as a critical consumer of data will become increasingly important.

Keeping track of public domain spatial data

Video, blogs, and online forums now provide many reference sources for spatial data. Listservs were one of the earliest ways to share information about public domain spatial data and many remain in use by the GIS user community today. ArcGIS Help has also emerged as a popular reference for learning about spatial data, particularly when users realized that the web version was better than the desktop version for two key reasons. First, the web-based version was updated much more frequently. Second, the online version links to the Esri Knowledge Base, which integrates discussions from around the world so that users can share their experiences and learn from each other.

Wikis are also beginning to play a role in informing people about public domain spatial data. Esri staff created the GIS Wiki in 2009 as a knowledge-sharing platform for better understanding of GIS concepts and improved GIS skills for increased productivity. All content ownership is shared by the GIS community, and, like other wikis, depends on the contributions of professionals, students, and the community-at-large to be successful. You can edit existing pages or add your own GIS-related content to the wiki. In a similar way, question-and-answer GIS user forums have, for several years, enabled the GIS community to grow. How will wikis and forums like these be monitored to ensure accuracy? It is generally easier to cast a critical eye on text responses than on mapped data, which often "appears" to be accurate. Most users appreciate that no one person has all of the answers, and that the answers posted aren't necessarily correct. Remember that wikis are, by definition, crowdsourced data with contributions from experts and non-experts alike. Again, take advantage of these new resources, but be cautious about the contents.

New methods and tools for accessing data

Not that long ago, users' choices of download tools were limited to choosing a file from a text-based list on an FTP site. As web portal technology evolved, users were increasingly offered the flexibility of downloading data from a map, first by drawing a box around an area of interest on a web map and downloading the data delineated by that box. This was the data access model for Esri's Geography Network, launched in 2000. Nowadays, data users might also be able to add layers from an online mapping service, customize the contents and appearance of those layers, then download the result. As we have seen, the model for downloading has also changed. In the past, users had to physically download the files; now those layers can be displayed as cached map tiles and accessed on the cloud.

Not only are datasets being shared more than ever before, but the tools that enable data access are also increasingly available. They are becoming more accessible, easier to use, and more functional. The APIs that now play such a big part in the GIS industry—Flex, JavaScript, and Silverlight—in just a few years have greatly improved in functionality and ease of use. HTML5 promises to provide the next generation of content delivery via the Internet. An ever-expanding repository of code snippets from Esri and elsewhere make it easy for developers to use and reuse code, much in the same way that HTML code has been copied from existing web pages to create new web pages.

It is likely that private industry will continue to play a key role in public domain spatial data. Several years ago, Esri developed a GIS Portal Toolkit which has evolved into an ArcGIS Geoportal extension. It is a toolkit of web-based and desktop software components that allow organizations to quickly establish geoportal sites to share data. It may be used to adapt the installation to a specific environment and create a host-specific look and feel for the interface. The extension was designed to enable end-users to discover geospatial data produced by others, preview those data, and make maps using a variety of map viewer technologies (such as Java, Flex, and Silverlight) that combine various geospatial resources. The portal also enables users to obtain that geospatial data, and the associated metadata records through a REST API that enables external access to these records.

Geoportals are based on the efficient storage and cataloging of metadata that are monitored automatically for changes, thus making spatial data searches much easier and more relevant for the end-user. They can also be configured to send out automatic notification of newly registered datasets that meet pre-defined criteria and expose an organization's own geospatial resources for others to discover. At the 2010 FOSS4G (Free and Open Source Software for Geospatial) Conference in Barcelona, Esri announced the release of the GeoServices REST Specification. This specification provides a standard way for web clients to communicate with GIS servers through REST technology. What does this mean for GIS users? As the specification is now open, developers will be able to embed geospatial services in their web-based applications, whether made available via ArcGIS for Server or non-Esri, back-end GIS servers or processors. At the same conference, Esri announced that the ArcGIS Geoportal extension would become open source. It was released in 2011 under one of the variants of the Creative Commons open source license and will include elements such as a geoportal web application, a cataloging service, a discovery service, metadata support, a configurable search engine, including a spatial ranking algorithm, federated searches to standards-based search providers, and publishing clients. Announcements like this from software vendors would have been inconceivable only a short time ago and reflect the increasing influence users exert on commercial organizations.

ArcGIS Online is also emerging, for a number of reasons, as a major force working to improve access to public domain spatial data. First, ArcGIS Online represents the best data from both the GIS user community and traditional data providers. Second, anyone working online with ArcGIS can take advantage of the services in the cloud, and have full access to all of their desktop GIS tools. Third, the ease of adding data to the ArcGIS portal through mapping services, layer packages, and map packages should ensure the repository of data will develop not only quickly but in a structured manner. Fourth, the ability for users to easily share what they have created with others means that it will become a recognized spatial data repository. Fifth, ArcGIS Online enables organizations to set up and easily maintain their own data portals customized to their own content. Sixth, the resource taps into the Amazon bank of servers, so the traditional constraints of files being too large to store in any one place are largely gone. For example, as previously mentioned, Landsat data spanning forty years has already been served online via ArcGIS. Users can examine land use and land cover change from the 1970s to the present easily and efficiently. This represents a significant change from the established workflows of the past twenty years, which involved searching for and downloading each Landsat scene required, making sure there was enough space locally to store it, reprojecting it if necessary, and symbolizing it appropriately.

A growing number of organizations (such as DataBasin, mentioned in chapter 9) are establishing an ArcGIS presence online and making even more data available to GIS users. In this decade, data portals will become more than portals; they will become communities.

Incorporating GIS into data services

As GIS joins the mainstream of information technology as an integral part of organizations' structure and workflow and as an expected output of government services, GIS public domain data will increasingly become an everyday component of government data provision. For example, the merging of federal statistics with maps through the United States' Data.gov portal was announced in 2010 (Sternstein 2010). Tools are being developed that will enable the public to easily create combinations of maps from the 270,000 datasets on the site. The first Chief Information Officer of the United States, Vivek Kundra, envisioned the website becoming an online marketplace where people worldwide can exchange entire databases and reuse content in ways the federal government could never imagine. This project is the result of collaboration between the EPA, the General Services Administration, the USGS, the US Department of Health and Human Services, and NASA. It reflects an increasingly common view that maps have an important role to play in making data understandable. It is, perhaps, a rather hackneyed expression, but a picture really does paint a thousand words—and a map is worth a thousand pictures. When Data.gov was made available online, Kundra reportedly said, "It's great that you have geospatial data in the catalog, but it doesn't mean anything to me if I can't see it." Mapping everything from mortality rates to houses with substandard plumbing is made possible by Geodata.gov, a separate catalog of geographic data that you learned about earlier in this book. Geodata.gov provides an invaluable extension to Data.gov by allowing you to view federal maps online or download them for further analysis.

Kundra said the next goal is to make the maps available as services or web-based mapping applications. The typical method of accessing spatial data is to download individual files and, for someone skilled in the use of a particular GIS software package, to manipulate and analyze those data. However, as you read in chapter 9, there is increased emphasis on providing access for the general public to what was formerly reserved for GIS users. If government data can be provided as map services, then anyone—policymakers to the general public—could visualize and analyze the data. These efforts support the entire GIS industry, as well. Private GIS companies can combine the data and maps with their software products to create custom applications for clients; others will be able to use the services to create and distribute free applications. It can transform government by providing greater transparency, more citizen participation, and increased collaboration between the public and private sectors.

As you read in chapter 8, resources such as a map-enabled Data.gov are turning maps into social phenomena in much the same way that Twitter and PicasaWeb turned short conversations and photographs into social objects. In chapter 9, you read that having the data on the Data.gov cloud paves the way for a two-way movement of data. Individuals will be able to contribute their own data, and maybe even broadcast their creations using Twitter, Facebook, or LinkedIn. Spatial data are becoming increasingly used in applications that have nothing to do with traditional GIS but still involve maps and spatial information. For example, Giuseppe Vasi's Grand Tour of Rome uses a geodatabase and website to explore the work of two eighteenth century masters of Roman topography, Nolli and Vasi. Sites like these blur the distinction between GIS and mainstream research and development. They establish a new audience to generate and work with spatial data and increase potential users who will demand services from and contribute to the GIS community.

New spatial data infrastructures

As governments develop plans for transportation, energy, and other vital areas of national infrastructure and grapple with critical environmental issues such as natural disasters and water availability, new national SDIs will continue to emerge. In many cases, these will develop in conjunction with e-Government initiatives and, to some extent, in response to continuing fiscal constraint as governments seek to maximize returns on their investments. A robust SDI is increasingly viewed as a critical foundation for more effective and efficient economic, physical, and social development. They are also viewed as essential for sound environmental management.

The Abu Dhabi Spatial Data Infrastructure Programme (*ArcNews* 2010) is an excellent example of just one such initiative. The Abu Dhabi SDI was established in 2006 on the heels of a massive e-Government initiative that was launched in 2005. By 2009, it was attracting three million visitors. This website contains everything from tourist information to land permits, from job postings to hospital locations. It also contains a personal document section for citizens, which includes land certificates, environment-related licenses, and even personal identification such as passports. The SDI is being developed in three phases, with the first to raise awareness among the government agencies about the relevance and goals of the SDI. The second phase involves the expansion of the SDI community to all relevant stakeholders, including over forty government and health organizations, businesses, and academic institutions. The third phase will involve adding methods and tools for cross-government geographic business intelligence and support, with the ultimate goal of a seamless geospatial network. The program eventually will include a comprehensive metadata catalog, policies, partnerships, standards, data, procedures, and technologies.

Through the Abu Dhabi geoportal, a data user can search for data, including images, map services, geographic datasets, geographic activities, spatial solutions, clearinghouses, and land information portals. Over 300 map layers were included by summer 2010. The layers can be searched by place name or via a map viewer, with searches further refined by the data theme, the format, or date. Full metadata can be viewed in the on-site map viewer or via GIS software. The site's map viewer also allows a user to save maps for later use, set transparency levels, query map information, and overlay map services. In keeping with our discussion on citizen science and cloud computing, users can provide geographic data by publishing map services and images, geographic datasets, geoservices, spatial solutions, geographic and land reference material, and even geographic activities or events to share with others through the submission of online forms.

To download data from the portal, the system currently asks for an e-mail address. Once registered, users will be notified when their data are available. The privacy policy on the portal assures visitors to the site that no personal information is stored about anyone who requests the data. As for guarantees about accuracy and currency, the site's authors make "no express or implied warranty or representation as to the accuracy, completeness, or up-to-date nature of the content or services on or accessed through this portal or any site linked to it or in respect of the services offered to the users." Neither the portal authors "nor any entity of the Government of the Emirate of Abu

Dhabi shall be liable for any loss or damage whatsoever arising out of accessing or using the content or services on or accessed through this portal or any site linked to it." The information on the site may only be downloaded, cached, displayed, printed, and reproduced if the following three conditions are met:

- The data may only be downloaded, stored, displayed, printed, or reproduced in unaltered form.
- The notice accompanying the data must be retained.
- The data can only be used for personal, noncommercial use, or use within an organization.

Data quality

More digital spatial data exist today than ever before. The sources of those data are varied in terms of producers, scale, format, purpose, attributes, geographic region, currency, completeness, and other characteristics. It is a multi-layered patchwork quilt that includes high-resolution satellite imagery; digital line graphs and other digitized topographic map content; nonmap-based digital data, such as remnants of old TIGER files; crowdsourced feeds and videos from current events; and data from sensor networks on the ground, under the oceans, underground, in the atmosphere, and from space. These are among many sources too numerous to list. As digital maps and databases become increasingly available, the quality of the data they contain become more varied. At one end of the spectrum, high-quality data from high-resolution sensors, sophisticated mapping techniques, and highly accurate GPS and other ground truthing methods continue to raise the bar of user expectations. At the other end of the spectrum, unverified data from crowdsourcing initiatives could mean data quality may be lower than ever before. In other words, good spatial data are getting better and lower-quality spatial data are getting worse, and there's more of it. As the amount of unverified spatial data eventually dwarfs authoritative data sources, the temptation to combine it with other layers will become more commonplace. This will have serious consequences for the built and natural environments and all those who inhabit them. On the other hand, the unverified data avalanche also provides a wealth of information that must be taken advantage of.

New developments are working to improve data quality. For example, as high-quality data are integrated into GIS, the errors in legacy data created years ago will become more apparent. Indeed, Nighbert (2010) illustrates the problem by overlaying hydrography, contours, and lidar imagery and observing the mismatches. Software advancements make it easier for organizations to make changes and view new data sources. For example, ArcGIS 10.1 greatly enhanced the ability for analysts to use lidar data. Given continued budgetary constraints, the easier and faster it is for organizations to make improvements, the more likely they will. Interestingly, as many governments place data online, they are performing a quality control check beforehand. It seems as though the prospect of showing one's data to the world is a powerful incentive to improve its quality. The cloud, therefore, could have a positive influence on the overall data quality of GIS databases worldwide.

For years, standards organizations stressed the need for data producers to provide metadata. But it wasn't until GIS software vendors made easy-to-use metadata creation tools available that good documentation became an established practice. Once it did, the catalogs and indexes that were created enabled the development of spatial data portals. In addition, tools such as Esri's ArcGIS Data Reviewer, which enables automation of the data quality control process, will make it even easier to improve data quality at its source.

As data users, you must always consider the quality of data and always review the metadata. With a better understanding of the nature of the data, you will make more informed decisions about how best to use the data.

Privacy, licensing, and copyright

In the rapidly changing landscape of GIS and public domain spatial data, issues of licensing and copyright are more important now than ever before. Technological trailblazing tends to precede legal safeguards. The development of GIS tools and the work of the geographic information user community have typically occurred at a much faster rate than the establishment of legislative frameworks governing the use of spatial data. The Creative Commons approach has emerged as a dominant influence, helping individuals and organizations obtain and publish the data that are increasingly in demand. Yet even in this environment and despite progress made, complications continually arise. One recent disturbing trend is the proliferation of different types of licenses for spatial data use, instead of a standard suite of licenses developed with the Creative Commons community. Some argue that a greater number of licenses only works to complicate and confuse the options.

With ArcGIS Online, Esri does not claim ownership of any content contributed by people or agencies; the data still belong to the contributors. Other repositories may have different philosophies and legal requirements. For example, at present every image created from Google Maps and Google Earth using satellite data provided by Google, through Digital Globe or other image provider, is a copyrighted document. Any derivative from Google Earth under US copyright law may be used only under the licenses Google provides. Google allows only noncommercial, personal use of the images, for example on a personal website or blog, as long as copyrights and attributions are preserved. By contrast, images created with NASA's World Wind SDK use Blue Marble, Landsat, or USGS images, each of which is in the public domain and may be modified, redistributed, and used for commercial purposes.

Issues of privacy will continue to be part of the landscape of public domain spatial data and GIS into the future. These issues will only increase in number and complexity as the number of high-resolution sensors increase while their physical size decreases. The protests over Google's Street View images that led to the company developing face-blurring technologies or, in other cases, completely blocking of the collection and publishing of street images are only the beginning of many such debates that will continue to simmer during the present decade. A few well-publicized and notorious cases of those who use spatial data for reprehensible objectives (such as property damage and loss of life) have raised awareness of the need for regulation.

Remember that maps are powerful sources of information and, as with all sources of power, can become embroiled in politics. The interaction among local, regional, state, and federal government agencies, international organizations, multinational

FOR FURTHER READING

To dig deeper into what the future might hold in terms of public information, copyright, patents, open source, Creative Commons, and more, have a look at this book by Lucie Guibault:
The Future of Public Domain: Identifying the Commons in Information Law (Kluwer Law International 2006).

The online book by David DiBiase, *Nature of Geographic Information: An Open Geospatial Textbook* (Pennsylvania State University 2008–2012), provides additional reflections on data, information, integration, and standards.

READ AND RESPOND

Read "How Creepy Does the Use of Geospatial Technology Need to Get Before There's a Major Backlash?" in Matt Ball's *Spatial Sustain* blog (Vector1 Media 2010). Ball wonders how "creepy," or how much privacy is invaded, when rapidly advancing spatial technology is combined with social media and handheld devices. He cites recent stories about a school district that was exploring the use of radio frequency identification (RFID) technology to track students, a landmark case in the US Ninth Circuit Court of Appeals that upholds the right of law enforcement to attach GPS tracking devices on vehicles parked on a suspect's property without obtaining a warrant, and a large full-body backscatter scanner that when mounted in a van can see through walls and into vehicles. What makes Ball most nervous are not the technologies themselves, but he sees the potential for a severe backlash that might affect less invasive sensor and location technologies that provide useful data. He is concerned that, in the name of profit, companies will create a whole new class of technologies that grossly violate rights and privacy and upset the market for the rest of the geospatial community. He thinks that the backlash against Google Street View, particularly after it was discovered that Google vehicles captured private Wi-Fi signals and some data, is only the beginning.

Do you think Ball should be concerned?

Should societies be more concerned about nefarious uses of spatial technologies by a few people intentionally seeking to do harm or about a backlash that could restrict the use of spatial technologies for constructive purposes?

Intertwined in these considerations is public domain spatial data—who produces it, how it is provided, how much is accessible to data users.

How could public domain data be affected by the possible backlash that Ball discusses?

corporations, GIS professionals, and ordinary citizens will continue to evolve and, at times, might become a battlefield of conflicting agendas and political objectives. The issues of regulation and open access will not go away. On the contrary, they will likely become even more contentious as the world works more closely together and realizes the power of GIS. The recent issue involving China's regulation of online mapping sites (Ball 2010), for example, is likely to be a recurrent hot topic in the years to come.

Free vs. fee data revisited

In chapter 4, we investigated some of the issues relating to whether government agencies should charge users for spatial data access. Given that this issue has been the subject of much intense debate over recent decades, you might be tempted to think that the point is moot. However, with many governments facing budgetary constraints and taxpayers reluctant to see their taxes raised for additional services or even to maintain current services, GIS-based services are increasingly viewed by local and regional governments as a potential revenue source. Some cities and counties offer a great number of online services, such as searchable and mappable property foreclosures, access to assessor's recorded documents, and even meeting minutes stretching back a century or more. Anticipating a backlash from local data users now accustomed to these data being provided for free, some local governments have proposed not charging for the spatial data but for access to value-added applications that have been developed in-house or outsourced to a private company. These applications require a significant amount of data processing to make them accessible and useful. For example, on top of the in-house data development and maintenance, one county in Colorado pays $25,000 per year to an outside vendor for hosting and developing their mapping site. The county could, if facing a shortfall in funding, discontinue that service and drop the GIS component, or it could ask the public to contribute to the project if it is perceived to be of value.

As you know from chapter 4, some argue that when governments provide data for free, they're not really free at all. It has taken a great deal of funding and resources (in terms of staff hours, software, computer hardware, and field equipment) to pay for data gathering, management, and hosting before those data could ever appear on a website. Providing data, in short, is never free, and simply maintaining existing services—never mind generating new and higher-resolution data—is expensive. People paid taxes to fund those services, state and federal funds were provided, and budgets were determined and authorized by strategic decision makers, such as county commissioners. In terms of the current shortfall in funding these data services, some point out that it was not the public's fault that a change in the economy resulted in less funding for specific departments in local government. Some state that new fees for data and services that the public already paid to develop are simply thinly disguised new taxes.

One benefit of free data is the increased efficiency in government and in other sectors that results from increased data usage. This would not be possible if imposing fees restricted use. In addition, government operations themselves become more streamlined as people monitor government activities through more open data policies. Others believe that fees to cover time and material costs are quite reasonable. By charging for services, they argue, users will have access to better quality data, and enjoy better performance and increased availability of data. Without fees, data users "get what they pay for," and the data quality may suffer.

Others believe that GIS technology provides so many benefits that they are willing to pay a fee for them. While the proportion of the population that is GIS aware has admittedly been small in the past, the future might turn out to be very different. The advent of citizen science, web mapping, and low-cost GPS-enabled devices could mean increased awareness of the benefits that GIS and spatial data can provide. Will this awareness lend itself to more support for the argument that all data should be free or will it lead to increased support for fees when those fees provide valued services?

Both sides might agree that, given the current fiscal situation, local authorities themselves should step up their efforts to attract funding from other sources. They could apply for grants from state and federal sources and consider public or higher education partnerships. Since these GIS services meet the needs of the community, some private funds might be available due to social concern and philanthropy. Another approach may be to separate services of interest to the public from services required by businesses, with the latter expected to pay for the services they use. If a business can charge money for the information they receive from local government, then many believe that business should pay for the government data services they use.

Some have set up regional GIS authorities to deal with the problem. These are usually quasi-governmental agencies, such as the Denver Regional Council of Governments or the Fremont County Colorado Regional GIS Authority. Most of the revenue is derived from the member organizations, such as counties, cities, school districts, fire districts, and sanitation districts. The members pay an annual fee and a small amount of revenue is derived from walk-in or call-in requests from property owners, real estate agents, developers, or others.

VGI and citizen science is prompting many to revisit the free versus fee debate. Van Oort and Bregt (2010) identified three influential developments: bottom-up free compilations of data, the falling costs of data distribution, and a preference for free data of inferior quality to fee-based data of higher quality. Restrictive pricing policies for public sector information could drive data users towards free online data. This may erode the user base and public support for tax-funded public sector information. They propose a mixed funding model that includes both fee and free data, depending on the "production phase" of the data user (development or production) and the type of user. One part of the proposal is that while a new public sector information dataset is under development, government or other prime users pay for access to those data, but once it is established in government work processes, access should become free for other users.

Whatever the future holds, it is likely that some local governments will continue to provide data for free, while others will charge thousands of dollars for the very same datasets. The debate will likely continue inside and outside of courtrooms. During 2010, in the case of *Sierra Club v. County of Orange*, a superior court judge in California ruled that a 640,000-parcel GIS database is considered software, not data. This is not just a question of semantics. The significance of the ruling is that, once it is considered software, the data are no longer subject to the state public records law which would require the local government GIS department to give its parcel database to the requesting organization, in this case, the Sierra Club, at the cost of reproduction. Now that the data are considered software, Orange County is allowed to demand a license fee of $375,000 for the parcel data. In his ruling, Judge James J. Di Cesare concluded,

> It seems undisputed that the Sierra Club can obtain a license from the County for use of the OC [Orange County] Land base in GIS format. It is apparent that issue comes down not to whether the OC Land base in GIS format will be made available but rather at what cost. While obtaining a license may be an expensive proposition, the County has shown that all of the revenue from its OC Land base in GIS format licensing accounts for only 26% of the costs to keep the OC Land base in GIS format up-to-date (Gakstatter 2010).

The Court of Appeal upheld Di Cesare's decision. Could the severe financial constraints facing many local governments bring renewed pressure that works against free and open data as governments look to their GIS assets as a source of revenue?

So much data and so many portals: Are all of them useful?

Having worked your way to the end of this book, you should appreciate that there are more spatial datasets and online portals now than ever before, with new ones appearing daily. However, along with the increase in the total number of available portals comes an increase in portals only for the general public and, therefore, portals of limited use to the spatial data user. Their use is limited for a variety of reasons. Most commonly, the maps they provide access to are not in the correct format; they may be PDF files, they may be unprojected image files, they may be graphics files such as Adobe

Illustrator, or they may be in some other non-GIS-friendly format. They may appear to be acceptable, but upon download and examination, they might contain errors, gaps, or worse.

Even the trend for more data produced as map services might not be useful for your project. For example, you read in the previous chapter that NRCS soils data are now online on ArcGIS. This has made a dataset, previously accessible only to the GIS community, available to everyone in an easy-to-use format with an ordinary web browser. However, at the current time, you cannot query it as a map service in the same way as you could if you had downloaded the same datasets into your desktop GIS. You can only either stream it as a background image and use the limited web identification tools. In the future, if data as mapping services are supported as input to geoprocessing and spatial statistics tools, the best of both the web and desktop worlds will truly be available.

With more and more people wanting access to spatial information and more and more spatially enabled applications appearing every week, you may be intrigued to read announcements in blogs and online news magazines about new data sources, only to discover that the format is not what you need. For example, a recent article announced that the USGS had just broken a record that had stood since 1972 by producing 2,500 new quadrangle maps over the past year (*Geomatics International Magazine* 2010). However, upon further investigation, the maps were stored as GeoPDFs, not currently easy-to-use in a GIS environment, and the data behind the maps were only partly new. Although the maps incorporated new NAIP imagery, the base layers were still sometimes decades old.

In the past, many data portals were often limited in scope to regional or national interests. Now, many spatial data portals are truly providing data at the international and global level; we have discussed several of them during the course of this book. For example, the World Database on Protected Areas is a comprehensive global spatial dataset on marine and terrestrial protected areas. It is jointly produced by the United Nations Environment Programme, the World Conservation Monitoring Centre, and the International Union for the Conservation of Nature's World Database on Protected Areas working with governments and collaborating NGOs. Through its portal you can download protected area boundaries and other data in shapefile and KML formats by using the advanced search tool on the website after registering as a user.

A simpler or more complicated world?

In many ways, life was simpler for data users in the early days of GIS. It was accepted that a large amount of the data would have to be created from scratch for most GIS-based projects during the 1980s and 1990s. As a result, no GIS lab at the time was complete without the ubiquitous scanners and digitizing tablets. Choices of base spatial data were limited, with the Digital Chart of the World's base data, Landsat satellite imagery, TIGER streets and boundaries, digital line graphs for hydrography and transportation, and digital orthophoto quads for aerial imagery providing most of the available information. Much data were either not in the public domain or only available at considerable cost to access and deliver. The difficulty of storing data on physical media, transporting the physical media, limited computing power, and the size of GIS data files largely confined GIS to specialized departments in universities, governments, and private companies. Although life seemed simpler then, life was not necessarily better for GIS users.

The good news, and also the bad news, is that now you have unprecedented choice in formats, projections, scales, and types of spatial data. You can stream data to your local computer, use data on the cloud, or download data through a variety of map- and nonmap-based portals. The challenge now is not so much about how to obtain and load your data into a GIS, but to understand which dataset is most appropriate to your needs. This process, far from being perfect, has become much easier for most core base data. As users of data, we are no longer content to settle for the best available data but strive to locate the most appropriate data.

Remember, just because a dataset is easily available does not mean that you have to use it in your project! Keep in mind the geographic inquiry process we referred to earlier:

- Establish the goals of your project.
- Focus on the problem you wish to solve.
- Only work with data that will help you solve that problem.

Then, like any project, weigh the cost and benefits in terms of both money and time spent on those data versus money and time spent on something else. The costs of converting data into the form that you need may well outweigh the benefits of doing so.

Data providers also have an unprecedented amount of choice. They can decide how much of their data holdings to make public and how they can best provide those data, whether on media or electronically, via a map service or via downloadable files, via tiles, as a seamless dataset, via free access or via a fee-based structure, with license restrictions or in the public domain. Increasingly, providers also must decide whether they will accept public contributions to the data.

The Association for Geographic Information (AGI) wrote a forecast of "The Geospatial World in 2015" (Coote et al. 2010). Among its predictions was that this world would include 120 GPS satellites orbiting, 320 location-based sensors would exist on 200 satellites, the mobile location market would be worth US$3.1 billion, smartphones would support over 10,000 separate location-based services, and all major United Kingdom conurbations would be mapped to an accuracy of 1 centimeter, through work commissioned by Google and Microsoft. Regardless of whether any of these things develop exactly as predicted, the geospatial world is undoubtedly heading into a decade of even more rapid changes. Public domain spatial data will be in greater demand, more varied, and more embedded in everyday decision making than ever before.

As a vast segment of society now has access to key spatial data layers and some online tools to explore them, spatial data will not only be used more often than ever before but will also be misused more often than ever before. In addition, modern web mapping tools mean that spatial data can be packaged, manipulated, and interpreted by an organization or agency, then hosted for the world to see. If you have the data yourself, you can more easily draw your own conclusions, for better or worse. If you rely on streamed pretty pictures, you are seeing the world through another's eyes. Knowing what they do about spatial data, GIS practitioners will most likely be called upon in this current decade to educate those who are using spatial data but are outside the GIS community in think-tanks, government agencies, in the news media, and elsewhere. You can play a key role in helping others make informed decisions and wise interpretations based upon spatial data.

READ AND RESPOND

Read the key challenges that Coote, Feldman, and McLaren wrote for AGI in the report, "The Geospatial World in 2015."

Of these challenges—political, economic, social, technological, and environmental—which do you think will have the most impact on public domain data? Why?

Mapping the world one volunteer at a time. Photo by Joseph Kerski.

In summary, you should now be aware of several significant trends in public domain spatial data. First, there are more public domain spatial data than ever before in the brief but amazing history of GIS. Second, accuracy and currency are improving in public domain spatial data to the point where real-time data on numerous phenomena are not just future predictions but present realities. Third, with the increase of volunteered geographic information and online maps that may alter and repackage original data, there is also a vast resource of undocumented, uncertified content in the cloud environment.

Given these trends, it is important that you become not just a consumer of spatial data, and one who appreciates their contents, but a *critical* consumer. A critical consumer understands the issues behind not only spatial data, but all data. A critical consumer is aware of the benefits *and* the limitations of spatial data, and understands the legal, ethical, and practical considerations of using spatial data to make decisions. Data users and data providers alike face the responsibilities and challenges to make the right choices for themselves and their organizations.

Recall your earlier investigation of the annual Esri Map Books. Now, examine the published maps while thinking about your newfound data skills, and ask yourself

Who made this map?
What skills did they need to have to assess, use, merge, and make decisions with data?
What skills do I lack that I need to develop?
How can I meet my goals more efficiently and successfully using spatial data?

It is clear that as GIS moves into the mainstream of decision making, communicating with GIS will be even more important and fundamental to understanding data. As Berry (2010) points out, we will need to use data to create "consensus maps" where data are used in an increasingly collaborative environment. The geospatial technology competency model approved by the US Department of Labor in 2010 makes it clear that technical competence is only one aspect of working with spatial data. Analysis, modeling, problem solving, decision making, creative thinking, personal effectiveness, academic and workplace competence, lifelong learning, planning, organizing, and so on, are all important.

The flood of information generated during the next ten years will make the vast amount of information available now seem almost negligible. If you, a discerning data user, are proficient at assessing and using data of all kinds, you will be well equipped for GIS and far beyond.

The future of GIS and public domain spatial data

During the first forty years of GIS development, people analyzed spatial data, both local and global, to make decisions. The next major step is to use spatial data to plan, and maybe even design, for the future. Besides hardware and software infrastructure, this requires a society that considers geographic knowledge fundamental to our sustainable future. The GeoDesign Summits held at Esri's headquarters starting in 2010 are a step in the right direction. The democratization of data—both their widespread use and creation—is creating a new geospatial infrastructure (Dangermond 2010). Hindering a full appreciation of the value of spatial analysis and decision making is an outdated

World topographic basemap captured as part of the Esri Community Maps project.

and low-technology educational system. Coupled with this deficiency is the lack of a curricular home for spatial analysis in the primary and secondary curriculum. (Milson, Demirci, and Kerski 2012) Even at the university level, geography, the traditional home for spatial analysis and GIS, is not as strong or as supported as it should be (Dobson 2007). Spatial analysis using GIS needs to expand beyond geography into business, health, mathematics, engineering, biology, and other programs (Gould 2010).

Despite great strides in recent years in terms of more open data access policies and new tools and incentives for data sharing, progress will remain slow unless certain things change. First, sharing data needs to be easier. Esri and data users accessing the ArcGIS Online resources, for example, have created templates to help people easily share the wide variety of datasets that they work with. This is one positive step forward. Second, publishing maps as shared services also needs to be easier. Advances planned in ArcGIS for Desktop to streamline the process of publishing layer and map packages from the desktop computer to a web mapping service without configuring a server will be a tremendous leap forward. Third, a greater amount of crowdsourced data needs to be authoritative data, from an agency or organization that understands how to create and document accurate and current spatial data. This is the focus of Esri's Community Maps Program, in which the governments and organizations that offer their data to the ArcGIS website are the ones that know the data best. They created the data and have been maintaining the data for years, sometimes decades.

Governments and organizations are the authoritative source of the data. Having access to accurate and current data is of critical importance, for example, in an emergency. At least some of the data used in crisis situations need to be from an emergency management agency and not from someone operating in a garage somewhere who maps as a hobby. However, increasing the percentage of online data from authoritative sources, or even making metadata in general easier to produce and maintain, will not guarantee that spatial data will be perfect. It will mean that you will be better able to make an informed decision whether to use those data or not.

Your vision for the future of GIS and public domain data

What are the big issues that you, the GIS community, will be asked to tackle during this decade? According to the Group on Earth Observations, a group of eighty-seven countries and the European Commission, our earth faces big challenges in nine societal areas, called the "strategic targets:"

- Disaster relief and response
- Health
- Energy
- Climate fluctuations
- Water conservation
- Weather systems
- Ecosystems
- Agriculture
- Biodiversity

Every one of these challenges has a spatial component that can be analyzed from a geographic perspective. The challenges also represent complex and overlapping issues;

for example, sustainable agriculture affects biodiversity, is dependent on a water source, and is affected by climate and weather. All these issues can be addressed more effectively with GIS as the primary toolkit. They can only be understood if they are visualized, and can only be visualized if they are represented by spatial data.

Goals identified by other organizations have similar core themes. These eight Millennium Development Goals have been adopted by all 192 United Nations member states and twenty-three international organizations in 2001 for implementation by 2015:

1. Eradicating extreme poverty and hunger
2. Achieving universal primary education
3. Promoting gender equality and empowering women
4. Reducing the child mortality rate
5. Improving maternal health
6. Combating HIV/AIDS, malaria, and other diseases
7. Ensuring environmental sustainability
8. Developing a global partnership for development

Every one of these goals has a spatial component and can be more effectively understood and grappled with using GIS, but time is running out. The availability and accessibility of spatial data are crucial.

Many of the organizations tackling these issues are nonprofit. They do not have corporate sponsors, large budgets, or adequate field staff. Most of the individuals using GIS within these organizations can only tackle these problems if there is supply of current and accurate public domain spatial data. Unfortunately, in many cases, data at a scale and level of detail appropriate to study are simply not available. To tackle these and other pressing issues effectively in the twenty-first century requires the continued development and availability of public domain spatial data and the placement of these datasets into the hands of researchers, developers, and policymakers. Easy-to-use yet powerful analytical tools are also required to allow those working with the data to tackle these complex issues themselves. Otherwise, public domain spatial data will be relegated to serve as backdrop images or layers, rather than data that can be interrogated, and from which new data and information can be derived.

Reflect upon and answer the following questions concerning public domain GIS data:

What would you like to see in the future landscape of public domain data?
Which datasets are missing from public domain data that you would like to see developed and made available?
Which scales, regions, and themes are missing?
Which attributes or classes are missing from existing data?

Which way for GIS and public domain data?. Photo by Joseph Kerski.

Today, as never before, data users have a voice in the entire process. Seemingly ordinary data users have done extraordinary things, ranging from creating and sharing open-source base data layers that have made government organizations wake up and take notice to advocating for the breaking down of some long-held traditions about how much data should cost and the restrictions on data use. Your role is no longer simply as an end-user: You can use data in a powerful way to create new, quality datasets and information. You can provide access to them for your organization's stakeholders or to a wider audience. You can contribute to local, regional, and global datasets started by others. You can influence data policies. By engaging in these activities, you can shape the future of GIS and public domain data.

References

"Abu Dhabi SDI Supports Wide-Ranging e-Government Programs Using GIS." 2010. *ArcNews* (Summer).

Ball, Matt. 2010. "How Creepy Does the Use of Geospatial Technology Need to Get Before There's a Major Backlash?" *V1 Magazine* (August 27). http://www.vector1media. com/spatialsustain/how-creepy-does-geospatial-technology-need-to-get-before-theres-a-major-backlash.html.

Ball, Matt. 2010. "When It Comes to Online Mapping Platforms, Is There More Than the Monetary Benefits of Local Search?" *V1 Magazine*. (July 2). http://www.vector

1media.com/spatialsustain/when-it-comes-to-online-mapping-platforms-is-there-more-than-the-monetary-benefits-of-local-search.html.

Berry, Joseph K. 2010. "A Brief History and Probable Future of Geotechnology." Paper distilling several keynotes, presentations, and papers. *Innovative GIS* and *Berry and Associates.* http://www.innovativegis.com/basis/Papers/Other/Geotechnology/ Geotechnology _history_future.htm.

Coote, Andrew, Steven Feldman, and Robin McLaren, eds. 2010. "AGI Foresight Study: The UK Geospatial Industry in 2015." *Association for Geographic Information.* http://www.agi.org.uk/ storage/AGI%20Foresight%20Study%20Summary%20Report%201.1.pdf.

Dangermond, Jack. 2010. "The Development of GIS—Yesterday, Today, and Beyond." Interview with Jameson Publishing and Field Technologies online president. *Field Technologies Online* (July 12). http://www.fieldtechnologiesonline.com/article.mvc/ The-Development-Of-GIS-Yesterday-Today-0002.

Dobson, Jerome E. 2007. "Bring Back Geography!" *ArcNews.* (Spring). http://www.esri.com/ news/arcnews/spring07articles/bring-back-geography-1of2.html.

Gakstatter, Eric. 2010. "Is It Software or Data?" *GPS World.* (May 25). http://www.gpsworld. com/gis/gss-weekly/is-it-software-or-data-9985.

Gould, Michael. 2010. "The Challenges of GIS Education Today." Interview with Jim Baumann. *GEOconnexion International Magazine* (November). http://www. geoconnexion.com/uploads/giseducation_intv9i10.pdf.

Milson, Andrew, Ali Demirci, and Joseph Kerski, eds. 2012. *International Perspectives on Teaching and Learning with GIS in Secondary Schools.* New York: Springer Science + Business Media B.V.

Montalbano, Elizabeth. 2010. "Gov 2.0 Summit Preview: GIS-Powered Government Data." *Information Week* (August 31). http://www.informationweek.com/news/government/ enterprise-apps/227200066.

Nighbert, Jeffrey S. 2010. "The Accuracy and Precision Revolution: What's Ahead for GIS?" *ArcUser* (Winter). http://www.Esri.com/news/arcuser/0110/accuracy-precision. html.

Sternstein, Aliya. 2010. "Data.gov's Next Big Thing: Mashing Up Federal Stats with Maps." *Nextgov* (June 18). http://www.nextgov.com/nextgov/ng_20100618_2467.php.

"USGS Mapping Productivity Record Slashed." 2010. *Geomatics International Magazine.* (September 21). http://www.gim-international.com/news/id5052-USGS_Mapping_ Productivity_Record_Slashed.html?utm_source=Newsletter&utm_medium=email &utm_campaign=20100921+GIM.

Van Oort, Pepun, and Arnold Bregt. 2010. "To Charge or Not to Charge: A framework for Proper Funding." *GIM International* 24(August):(8) 13–15. http://www.gim-international.com/issues/articles/id1572-To_Charge_or_Not_to_Charge.html.

Glossary

application programming interface (API): A set of interfaces, methods, protocols, and tools that application developers use to build or customize a software program. APIs make it easier to develop a program by providing building blocks of prewritten, tested, and documented code that are incorporated into the new program. APIs can be built for any programming language.

array: A set of objects connected to function as a unit. In GPS technology, an array of satellites that is used to pinpoint locations on the earth.

broadband: A high-speed data transmission network connection allowing large amounts of information, such as data, graphics, or video, to be transferred quickly.

catchment: The area of land from which water drains into a reservoir, pond, lake, river, or stream system.

cloud computing: Cloud computing furnishes technological capabilities—commonly maintained off premises, and not on the local computer—that are delivered on demand as services via the Internet.

conductivity: The ability to conduct, or transmit, an electric current. In water quality, conductivity provides a measure of the purity of water—the lower the conductivity, the purer the water.

data dictionary: A catalog or table containing information about the datasets stored in a database. In a GIS, a data dictionary might contain the full names of attributes, meanings of codes, scale of source data, accuracy of locations, and map projections used.

DPI: Initialism for *dots per inch*. A measure of the resolution of scanners, printers, and graphic displays. The more dots per inch, the more detail can be displayed in an image.

electromagnetic spectrum: The entire range of wavelengths (or, the range of frequencies) over which electromagnetic radiation extends.

flow accumulation: Defines the amount of upstream area draining into each raster cell.

flow lines: The path describing a moving particle of water.

geocode: A code representing the location of an object, such as an address, a census tract, a postal code, or x,y coordinates, such as latitude and longitude.

geoprocessing: A GIS-based operation or technique that manipulates a set of spatial data. Geoprocessing takes an input dataset, performs a specific operation on those data such as proximity or overlay analysis, and returns the result of the operation as an output dataset.

georectify: The digital alignment of a satellite or aerial image with a map of the same area. In georectification, a number of corresponding control points, such as street intersections, are marked on both the image and the map. These locations become reference points in the subsequent processing of the image.

GPS: Initialism for *Global Positioning System*. A system of radio-emitting and receiving satellites used for determining positions on the earth. The orbiting satellites transmit signals that allow a GPS receiver anywhere on earth to calculate its own location through triangulation. Developed and operated by the US Department of Defense, the system is used in navigation, mapping, surveying, and other applications in which precise positioning is necessary.

headwater: The source of a river or stream.

HTML: Initialism for *Hypertext Markup Language*. A computer language used to create web pages for publication on the Internet. HTML is a system of tags that define the function of text, graphics, sound, and video within a document. Now an Internet standard maintained by the World Wide Web Consortium (W3C).

intellectual property rights: Defines the rights and privileges attached to ownership of intellectual property (any form of original creation that can be bought or sold).

KML: Initialism for *Keyhole Markup Language*. A file format, based on an XML schema, for displaying geographic data in 2- and 3-dimensions via web browsers, mapping visualization software, and GIS.

lossless: Data compression that has the ability to store data without changing any of the values, but is only able to compress the data at a low ratio (typically 2:1 or 3:1). In GIS, lossless compression is often used to compress raster data when the pixel values of the raster will be used for analysis or deriving other data products.

lossy: Data compression that provides high compression ratios (for example 10:1 to 100:1), but does not retain all the information in the data. In GIS, lossy compression is used to compress raster datasets that will be used as background images, but is not suitable for raster datasets used for analysis or deriving other data products.

map cache: A setting used in ArcMap that allows temporary storage of geodatabase or online map service features from a given map extent in the desktop computer's RAM, which might result in performance improvements in ArcMap for editing, feature rendering, and labeling.

metadata: Information that describes the content, quality, condition, origin, and other characteristics of data or other pieces of information. Metadata for spatial data may describe and document subject matter; how, when, where, and by whom the data were collected; availability and distribution information; projection, scale, resolution, and accuracy; and data reliability with regard to some standard. Metadata consist of properties and documentation. Properties are derived from the data source (for example, the coordinate system and projection of the data), while documentation is entered by a person (for example, keywords used to describe the data).

middleware: Software packages or programming that connects application software and operating systems, allowing multiple processes running on one or more machines to interact.

orthophotograph: An aerial photograph from which distortions owing to camera tilt and ground relief have been removed. An orthophotograph has the same scale throughout and can be used as a map.

orthophotoquads: An orthophotograph that has been formatted as a US Geological Survey (USGS) 1:24,000 topographic quadrangle with little or no cartographic enhancement.

orthorectified: The process of correcting the geometry of an image so that it appears as though each pixel were acquired from directly overhead. Orthorectification uses elevation data to correct terrain distortion in aerial or satellite imagery.

pixel: The smallest element of a display device, such as a video monitor, that can be independently assigned attributes, such as color and intensity. Pixel is an abbreviation for picture element.

planar: A two-dimensional measurement system that locates features on a plane based on their distance from an origin (0,0) along two perpendicular axes.

quadrangles: A rectangular map bounded by lines of latitude and longitude, often a map sheet in either the 7.5-minute or 15-minute series published by the USGS. Quadrangles are also called topographic maps or topo sheets.

scale: The ratio or relationship between a distance or area on a map and the corresponding distance or area on the ground, commonly expressed as a fraction or ratio. A map scale of 1/100,000 or 1:100,000 means that one unit of measure on the map equals 100,000 of the same unit on the earth.

schema: The structure or design of a database or database object, such as a table, view, index, stored procedure, or trigger. In a relational database, the schema defines the tables, the fields in each table, the relationships between fields and tables, and the grouping of objects within the database. Schemas are generally documented in a data dictionary. A database schema provides a logical classification of database objects.

spatial analysis: Modeling and investigating local to global phenomena, problems, or issues using the geographic perspective in conjunction with GIS, spatial statistics, and other techniques to understand patterns, relationships, and trends across space.

stereo compilation: A map produced with a stereoscopic plotter using aerial photographs and geodetic control data.

stereo correlation: The extraction of three-dimensional measurements from a stereo pair of photographs.

topology: In geodatabases, the arrangement that constrains how point, line, and polygon features share geometry. For example, street centerlines and census blocks share geometry, and adjacent soil polygons share geometry. Topology defines and enforces data integrity rules (for example, there should be no gaps between polygons). It supports topological relationship queries and navigation (for example, navigating feature adjacency or connectivity), supports sophisticated editing tools, and allows feature construction from unstructured geometry (for example, constructing polygons from lines).

toponymic: Related to the study of place names.

URL: Initialism for *uniform resource locator*. A standard format for the addresses of websites. A URL may look like this: http://www.esri.com. The first part of the address indicates what protocol to use (such as http: or ftp:), while the second part specifies the IP address or the host name (including the domain name) where the website is located. An optional third part may specify the path to a specific file or resource (http://www.esri.com/products.html).

vertex/vertices: One of a set of ordered x,y coordinate pairs that defines the shape of a line or polygon feature.

volunteered geographic information: Spatial data contributed by individuals, rather than traditional mapping organizations, who create, gather, assemble, and disseminate via the cloud environment. These individuals are typically not paid by an organization.

wavelengths: The distance between two successive crests on a wave, calculated as the velocity of the wave divided by its frequency.

XML: Initialism for *Extensible Markup Language*. Developed by the W3C, a standardized general-purpose markup language for designing text formats that facilitates the interchange of data between computer applications. XML is a set of rules for creating standard information formats using customized tags and sharing both the format and the data across applications.

Index